THE ORIGINS
OF SELF

By Dr. Stephen J. Brewer

www.originsofself.com

For Anne, our children and grandchildren

ISBN-13: 978-1484080931
ISBN-10: 1484080939
A Stephen J. Brewer publication.
Printed by CreateSpace Independent Publishing Platform

Contents

Acknowledgments:

I would like to thank the many friends, family, colleagues and members of online forums who through debate and discussion helped me to clarify the arguments developed in this book. I give special thanks to my daughter, Katharine for her editorial suggestions and proof reading.

Preface

When it comes to important decision-making, I certainly feel I am the master of my body and the cause of all the actions I take. Science does not agree; instead, it has produce powerful evidence showing that human beings are machines programmed by algorithms designed to ensure its survival and reproduction. All decisions are made by these algorithms. This 'me' believing it is in charge and able to make free decisions is just an illusion produced as a side effect of the machine's operation.

I believe this conclusion is not just nonsense, but dangerous nonsense. It is dangerous because the entire structure of our civilization depends on me being a free moral agent responsible for my actions. If I am not free to act, then I am not accountable for my actions. I argue to the judge, 'It wasn't me that did it, it was my brain'. Furthermore, every time we describe a human as a mere machine, we leave ourselves open to the most destructive elements of our nature. This is because deep down I know I am real with a powerful will, but it seems I can do what I like to another person because this other is just a machine with the illusion of self. Therefore, 'it' can't actually suffer. Many horrors against humanity have been committed when individuals are indoctrinated into believing another person is not a 'real' person. Our machine-like vision of humanity currently pervading the media is in danger of leading us once more down this fearful path.

I believe the science leading to the 'consciousness is an illusion of a machine' is nonsense because it was obtained by treating life as if it were just a physical object. This is a disaster when trying to understand how life responds to external forces. This purely physical treatment ignores the obvious fact that life actually experiences and uses these forces for its own ends. If these experiences are not to be mysteriously derived, then they need to be part of the physical world itself. Such physical feelings then allow animal consciousness to become causally linked to the physical world. This means that our consciousness is no longer an illusion or a mystery, but an integral component of our body's functioning. It has helped to make our bodies what they are, and do what they do.

How could I, an industrial bio-scientist, dare to argue that our physics is incomplete? The story starts with my need to direct a multidisciplinary team of forty scientists. Such a team is typical of the mixture of skills needed to produce modern medicines. I could not be a master of their expertise but I could understand the general principles lying behind science itself. This naturally lead me into the fascinating world of the philosophy of science.

A side effect of this study was the reopening of my curiosity about life and the nature of being. It was during these studies that I came across Alfred North Whitehead's 'Process and Reality' (1929). Although the first reading went way over my head, those parts I did

Stephen J. Brewer

grasp and his sub-title 'philosophy of organism' made me determined to understand this work. This, as it turned out, set me on a 20-year task to read all his various references to other philosophers, as well as subsequent contributions to the philosophy of mind. Each cycle of re-reading 'Process and Reality' against this ever-broadening philosophical background convinced me his work could make a major contribution to our understanding of life-science, evolution, mind and the nature of the conscious self. But most importantly, its powerful arguments could heal the rift caused by these self-destructive pronouncements that alienate science from society.

The problem was how to communicate Whitehead's discoveries. My intended audience is not only scientists but also members of the public concerned about the way science is undermining their humanity. I also had to ensure philosophers would feel I had a good grasp of this notoriously difficult metaphysics. This meant I needed to cover both the technical and the philosophical concepts required to understand the true nature of conscious living organisms. The result is this surprisingly short book! It is short because I want to communicate these findings concisely but accurately to busy people. Therefore, I have not attempted to reproduce the enormous scope and subtle arguments Whitehead uses to join his concepts into a beautiful coherent whole. Instead, my book explores only those parts relevant to the debate about the origins and nature of the conscious self.

You may be surprised to see the book is not in the form of a lecture, but is a series of dialogues. This is because to understand Whitehead's work, I had to test my half-formed (and often half-baked) ideas on fellow scientists, friends and relations. To do this I always needed to start my arguments from first principles. Only from this foundation could we explore some of the bigger concepts. As a result, it seems natural to also write this book as a series of conversations and arguments. This will allow you to repeat my process of discovering the many concepts of this philosophy from these very same first principles. If you do, I promise you will see the world and our place in it from a radically different but highly enlightened perspective.

Finally, in order to maintain the flow I restrained my natural instinct to insert footnotes and references. Instead, at www.originsofself.com my readers will find an appendix containing my notes related to the underlined text. The appendix includes links to a wide selection of authoritative published articles (mainly open access) forming the foundations for the radical concepts outlined in this book. It is also here where I invite readers to join this debate by leaving their own thoughts, comments and corrections.

Introduction

I have obviously derived the title of this book from Darwin's 'The Origin of Species'. I have done this because my book is a development of his theory to explain not only the evolution of bodies but also the co-evolution of our minds and consciousness.

At first reading, the concept of animal consciousness appears built into the Darwinian 'struggle for survival'. This phrase seems to imply the presence of a person, a struggling self that really cares about whether or not it survives. The existence of a struggling self, an animal person, was a matter of considerable debate during the latter part of the 19th century. Darwin, at least initially, sided with the argument that all animals really struggle. The big problem with this concept occurred with the discovery of a vast underlying layer of life in the form of microscopic single celled organisms. To give personhood to such primitive forms of life seems such a ludicrous concept that by the end of the 19th century, the struggle for survival is described in terms of the animal's disinterested and mechanistic reaction to environmental stresses.

During the next century, biochemists revealed the mechanisms responsible for the functioning of these microscopic organisms. They are composed of a complex system of molecular scale machines able to process and produce all the chemicals required for life. All this occurs without violating any of the well established laws of physics and chemistry. In the middle of the 20th century, DNA was shown to be the main carrier of genetic information. The complex mechanisms allowing this information to be transcribed into these functional mechanical units was then elucidated. With the unification of biochemistry and genetics into a grand synthesis, we saw the dawn of biotechnology and the ability to determine our own genetic future.

One hundred and fifty years after Darwin's publication and with all these advances in life sciences, there is only one problem left, we still cannot explain how this machine can also be conscious. But why bother? After all, we can explain the evolution of these machines without a subjective struggle for survival, so why not explain away this 'self' as being an illusion somehow emerging from the shear complexity of this living machine. This gives rise to the concept that the self is a form of <u>narrative</u>. Free will is that part of the story where the self retrospectively justifies actions over which in fact, it has no control. The problem with this solution is that even an illusion of self needs to be rooted in something. Therefore, we are still left with the problem of finding the origins of self.

Against this backdrop, Freya, a university professor and molecular biologist, dares to question the dogma that life can evolve without the presence of an aware subject. Max, a biochemical engineer, finds himself vigorously defending the conventional view where evolution is entirely explained by a combination of chance mutation and natural selection with no need for animals to be in any sense aware. Part I of this book focuses on this debate. As the arguments continue, they both begin to understand how an embodied chemical mind must be present in even the simplest of living forms. Encouraged by this, Freya argues this mind must be aware of events in its environment. Max then shows Freya that if this were so, the very chemicals used to form living organisms must also be aware. Since as a scientist she cannot support such a ridiculous conclusion, she must concede to Max's powerful argument. The only possible explanation for a conscious self is that it is an secondary effect only emerging from living systems with a high level of complexity.

Part I establishes the technical basis and philosophical issues to be developed in Part II. This is where Orin, a philosopher-scientist and Freya consider how physical events, such as the adsorption of a photon of light energy, would be experienced by simple chemical systems. They show that without changing any laws of physics, such events could be the origin of an animal's awareness of its environment. They also see how an animal's will-for-life originates from the actions of its self-reproductive chemistry. When such an unstoppable will drives this aware embodied chemical mind, then even the simplest of animals will really struggle for survival. As their dialogue continues, they show how the adoption of these concepts fully explains the evolution of animals with consciousness, self-consciousness and free will. During this dialogue, Orin and Freya also produce new insights into the essential nature of self within the functioning of our own incredibly complex embodied minds. Finally, they explore how self-consciousness allows humans to develop language, religion, reason, civilization and scientific knowledge.

In part III, Orin shows how this new understanding of self incorporates many aspects of the process philosophy developed by A. N. Whitehead during the early 20th Century.

I hope you can join in this adventure of debate and discovery to re-establish the primacy of our personhood while fully recognizing our inter-relationship and growth from the natural world. You are most welcome to join this debate by leaving your thoughts and comments at www.originsofself.com Here, you will also find an appendix with explanatory notes relating to the underlined text as well as a substantial list of references.

Part I: The Debate

Chapter 1: The Missing Me

In which Freya reveals her problems with the Theory of Evolution and Max determines to expose the flaws in her reasoning process.

All about me

Freya: I've finally lost my faith in Darwin's Theory of Evolution!

Max: O my God, you've become a Creationist! You go to my family reunion, meet my fundamentalist cousins and they've won you over. That's despite all my brilliant attempts to argue them out of such nonsense.

Freya: You made your point very powerfully, as you say, it's not a 'theory' that animals evolved from simpler organisms over many millions of years, it's a matter of factual evidence based on the fossil record.

Max: Why they need to turn these facts on their head to justify some ancient cosmology is what I don't understand. We seem to be living on an entirely different planet. They're suspicious of all science, not just evolution, but global warming, genetic engineering and particle physics. As far as they're concerned it's all a left wing conspiracy to undermine their God, freedom and the American way of life.

Freya: Well your all out attack didn't help to bridge the gap either. All their stereotypes of scientists being left wing amoral atheists were confirmed weren't they?

Max: Well perhaps I did get a bit carried away. So what's this ridiculous statement about you 'losing your faith'?

Freya: I still believe evolution happened, but not in the way described. Something in Darwin's theory just doesn't work.

Max: Oh I see where this is going; it's the process of natural selection you don't like. This is just the bedrock of evolutionary theory. If you're going to challenge that, then you have lost your head totally. To even contemplate such a claim means you're going to take on some mighty big guns of the scientific establishment. It's goodbye to your university career and tenureship. Good job I've got a decent income, we biochemical engineers don't have the time to ask such questions, we just get on with the job of making the medicines that cure cancer.

Freya: You see my problem with the theory is a personal one. It's about us, or rather me. We don't come into the picture at all, I've no role to play in what happens so what am I for?

Max: Me, me, me, it's always about you. Seriously though, it's quite obvious, I'm the one that struggles for survival and so do you. It's because I struggle to survive I bring home the food so we can bring up our children. Without my struggle, we wouldn't survive and neither would our family. My ancestors passed on my genes to me because they struggled and survived. If they didn't they would have died out and I would never have been born. My genes are fit for survival so I'm fit for it as well. You can see that just by looking at my perfectly symmetrical body! I can survive whatever nature throws at me out here in the back woods, well the suburbs anyway.

Freya: That's just where you're wrong and fallen into the trap most people fall into when they hear about 'the struggle for survival'. We don't struggle at all, or at least our evolutionary ancestors, monkeys, fish, plants etc. never did.

Max: Don't be silly, there's an entire chapter in Darwin's 'The Origin of Species', all about how I have to struggle to survive.

Freya: But have you ever actually read it? No, you like most people haven't, but if you did you would see it's not *your* struggle or *our* struggle at all, its 'The Struggle'. It doesn't belong to you; it's all to do with the environment and the limited supply of food. It's like the force of gravity, that's not your force is it? It's a force, out there entirely indifferent to you as a person. That's just as it is for the Struggle for Survival.

Max: Yes I sort of know that, but all the same, it does seem as if animals actually do struggle. Unlike rocks, they want to survive. It's just those with the best genes are most likely to survive by running faster than the predators can. They want to run away because they feel fear. I've seen it in them and felt fear myself.

Freya: Well at the end of the 'Struggle' chapter, Darwin lets us know 'no fear is felt' by the animals. I know it seems to contradict everything we think we know about animals, but apparently he seems to be of the opinion they just act as if they are fearful.

Max: If a bear came after me, I'd definitely feel fear, and run like hell to escape, or I would turn and fight, pick up a branch and hit him on the nose. I'd at least go down fighting. My 'fight or flight' genes would take care of me. So is he saying with animals it's different?

Freya: Well, that's where it becomes a bit confusing. For example Darwin also says 'in this war of nature, the vigorous, the healthy, and the happy survive and multiply'. It's odd to say no fear is felt, and yet the animals are happy. Surely happiness is just as much a subjective state as fear; so why should one be felt but not the other?

Max: He seems as confused as we are. He must mean something else than them having an emotional state of happiness, although I don't know what that could be. Perhaps he just wanted to make you feel happy so you're not depressed thinking about all the animals killing and eating each other in the name of evolution.

Freya: All the same, I think as a naturalist, especially when dealing with highly advanced animals, it must be impossible not to recognize you're looking at another 'person' experiencing similar types of <u>emotions</u> to us. The concept of a 'war of nature' is yet another very subjective term. There can be no war for entities not seeking fulfillment of some subjective need. They are just robots battling it out in a world where nothing cares about the results anyway. In his next book the 'Descent of Man', Darwin discusses animal intelligence and decides even lowly animals do have feelings and actually struggle to get free from a predator, and enough intelligence to work out how best to do this. So, during the late 19th century all sorts of discussions took place about the extent to which animals are intelligent, conscious and feel emotions.

No one struggles

Freya: Anyway, it was finally argued naturalists had just projected their own feelings on these animals, so even if they behaved as if they were intelligent they weren't actually conscious. It's just something called instinct, some automatic system. To cut a long story short, the conclusion was it doesn't matter to evolution whether simple or for that matter complex animals were actually intelligent and have real feelings and were conscious, let alone self-conscious. Animals would be treated as automatons. Since then the theory of evolution has developed very happily without any real reference to a subject actually struggling for survival. Subjectivity and 'real' conscious intelligence is seen as emerging as a result of evolution, somehow appearing when the organism gets a large enough brain and needs to live in complex societies. There's no role for a feeling intelligent subject in the evolutionary process at all.

Max: I can live with that. If it's no problem for them, why should it be one for you?

Freya: The problem is that something very odd is going on here. If you deny the presence of an actual struggling subject in the animal, a subject caring about itself, then you have to deny its presence in us, or accept we really are something new and special. We would then indeed have a special relationship with the world, just like it says in the Bible.

Max: To think we are something new and special goes against the grain of evolutionary science and the elimination of God from the whole process.

Freya: That's why the neo-Darwinists like Dawkins come to the very honest but highly disturbing conclusion we are all 'lumbering robots', machines whose sole purpose is to ensure our genes survive. They call all the shots; we're simply programmed to think it's us doing it. We have no will of our own; we just do what our genes tell us to do. Our subjective self-determination is an illusion, of no consequence to evolution at all. In fact taking it to the extreme we are not of any consequence at all. This is the only logical conclusion you can make if you follow the 'self is an illusion' line of reasoning.

Max: That's it then, there is no consequence, no reason and no purpose. I can behave as I like with no real consequences because you and they don't exist anyway.

Freya: Yes, the very essence of what we like to think makes us human and keeps us civilized is based on some illusion of a caring self-determining person being present. No wonder this line of thought causes such an anti-science reaction. Far from Creationism being beyond reason, many see it as the only way open for people to defend their humanity against such a totally dehumanizing science.

Taking part

Freya: So we're going to get to the bottom of this and find out whether or not you and I are mere ornaments on the process of evolution, or if we, or some subjective component that really cares has been a driving presence throughout it all. Here's my argument: I intuitively feel that something like 'me' is present in all animals. Sort of enjoys the experience of living, it actually cares, and because of that it has a key role to play in evolution because it really does struggle to survive.

Max: So, just let me get my head around this. You say the phrase 'struggle for survival' is describing exterior environmental forces acting impartially on every living thing, just like the force of gravity. You're proposing we need to find an interior force of something like 'care', or a will-for-life. You say we need the presence of an interior subject, a person actually caring about how it's affected by this blind exterior force.

Freya: Yes, the total opposite to how stones don't care when acted on by the force of gravity. A sort of internal resistance to the blind forces of nature if you like.

Max: But you need to accept a caring 'I' might in fact be a very recent evolutionary development, one only humans and perhaps some close relatives have. It might be a real aid to our survival. Perhaps it even accelerated evolution of animals caring about their surviva_. After all, we've broken free of this blind force by being able to use genetic engineering and medicines to overcome genetic deficiencies. But it still means most of evolution didn't have this internal caring self to struggle against the external environment. Maybe all higher animals do care; fish even, but what about plants. We say they 'struggle for the light', but we don't actually think they have a subjective state of yearning.

It gets even worse, what about the subject of your research and the things I used to grow in my lab, bacteria the very simplest of life forms. They were the very first forms of life and will probably be the last to die out. Where would a caring self be in these tiny bags full of proteins and DNA?

Freya: OK, you've made your point, and I think you've got to the reason why the caring self had to be ignored if Darwin was to produce a general theory of evolution. Although Darwin might actually have believed a 'self' was present in the more advanced animals, arguing lower animals also possessed some form of caring self would make him a laughing stock of the real scientists of the time. Even worse, it might suggest animals have souls, offending the religious establishment even more than he was about to do anyway!

Max: Then as now, physics and its treatment of inanimate objects is seen as 'real science'. Biologists are just a bunch of lightweights.

Freya: Yes, and if biology was to reach a similar

stature, then any science of living things had to match up to physics. Life science didn't want to open the 'can-of-worms' by introducing the person, the 'I' into the things it studied. Biologists didn't want to introduce the soul into the machine as it were. Even if 'care' does develop at some latter stage, perhaps acting as a booster rocket to the evolutionary process, Darwin and the neo-Darwinists had to ignore it to make it a general theory for all life forms.

Max: Since then, the main stream of evolutionary science has done very well without it. So the conclusion: we're some late arrival, an ornament on the process, but unnecessary to the whole flow of evolution; case closed.

Becoming complex

Freya: Well you see it's not quite that simple, and now I'm really going to rock the boat. The neo-Darwinists say they have evolution all explained, but the truth is they don't actually explain evolution itself, not at all! If natural selection is a purely mechanistic and physical process operating on inanimate objects, how could it cause complex organisms to evolve? I agree it explains how when the environment changes, those animals with the best genetic survival resources will be selected. It does this without the need for any self, but evolution has also got to explain why animals became so complex over time. That's exactly what the fossil record has shown to happen. Natural selection is all about accessing the solutions already present in a vast pool of genes. It's about adaption to the environment and this is often achieved by mutating control genes. These are the ones responsible for turning on whole banks of other genes making bodily structures such as a bird's beak. Run this set of genes a few more times and you get a longer beak, the same for legs and necks. It's all occurring by subtle mutational changes at a very sophisticated level of control. Present day animals are primed for adaption to new environments. The mixing of genes by sex is the key to making sure somewhere in the population there will be offspring able to make the transition to a new environment. If not, it's the end of that particular species, but the massive gene pool lives on, this time in other creatures.

Real evolution is how we start with the primordial soup of chemicals; then how simple organisms emerge like bacteria; then the more complex cells, like the amoeba, and algae; then multi-cellular life, worms, etc. The simple animals came before complex ones. What drove that enormous increase in complexity?

Max: Well, that's all worked out isn't it. The more complex the entity is, the better it's able to compete. It would be more adaptable, make a wider range of more appropriate responses to changes in the environment.

Freya: Since any such 'appropriate response' has got to be a 'reasonable' one, then this means the growth of some form of internal intelligence along with increased complexity. So we aren't talking about the evolution of inanimate objects at all, but about entities possessing and using their own intelligence to reproduce and survive. Surely, if we have such an intelligence, then we must have an entity with a very primitive form of awareness?

Machine intelligence

Max: I don't necessarily agree. If it's a purely computational type of response, there's no aware intelligence acting. My computer does calculations and makes logical decisions, but it's certainly not self-aware. And although the artificial intelligence guys claim one day it may be so, it's very far from being so at the moment.

Freya: Look, I don't mean simple organisms are going to be self-aware, either. Certainly not in the way we are self-conscious or even the way other advanced animals such as cats and dogs are conscious of the world. What I mean is that it's able to experience its environment in a very limited way. For example, it will respond to the presence of food, selecting and processing only those chemicals serving its reproductive purpose and avoiding those that don't.

Your computers can't reproduce themselves, or compete with others for food by searching out power plugs. It just shows computer 'intelligence' is one of the lowest grade analogies you could use if you want to approach the intelligence shown by even the simplest of bacteria!

Max: Well let's just call these reactions to the environment instinct, and leave it at that.

Freya: Now you know that's just a cop-out. 'Just instinctive' is no different from saying 'just intelligent'. Instinctive behavior is the outcome of some underlying intelligence and that's what we're after. After all, even the humblest of life forms respond in a very logical way to changes in their environment. Bacteria are able to stay alive on a wide range of different foods and some are able to survive in extreme environments. They're also pretty successful at it, since they've managed to enter into the very deepest recesses of the planet as well as able to survive in outer space in the form of spores; just waiting for a bit of fertile ground, and off they go again.

It looks like we've already run head-on into what we actually mean by the word 'intelligence', or 'instinctive' or 'logical behavior'. How can we search for the origins of self in these simple organisms, if we can't even agree on what we mean by the words 'intelligence' and 'instinct'?

Max: Just to show the futility of your quest for the origins of this caring self, this entity you claim is needed to provide a real struggle for survival, you first need to justify how you can ever call bacteria 'intelligent'.

Chapter 2: Evolution of Mind

In which Max and Freya discover how a simple chemical reaction caused the evolution of highly sophisticated embodied chemical minds able to survive and reproduce.

Creative chemistry

Max: The simple fact is science can explain the evolution of life using pure chemistry. We can sketch out all the chemical steps, from the 'primordial soup' of organic chemicals formed by lightning strikes in earth's primitive atmosphere, right up to the emergence of the first life forms.

To turn this primordial soup of organic chemicals into life we just need a relatively simple self-replicating chemical. Crucially, chains of RNA molecules, the simple chemical found in all living cells, can catalyze their own production. There are four slightly different variations of RNA, which means you can make a chain of only ten RNA molecules a million different ways.

Now, some of these RNA based auto-catalysts are better at making copies of themselves than others. This means you have all the conditions needed for the Darwinian process of evolution. You have competition for the RNA 'food' needed to make the copies and the generation of variants, some of which will be better than other ones. For example, the first variant able to combine three chains to make a copy would be able to exploit a new source of 'food'. Another example would be the ability to distinguish useful 'food' from 'poison'. These poisons would be very sticky RNA chains binding so tightly it slows or even stops the copying process. The concept is any variant possessing more discriminating and faster chemistry will out-compete others.

This pre-biotic stage of evolution is likely to occur in tiny niches amongst crystalline crevices in rocks where the chemical conditions are just right. Now another key component of cellular life, a cell membrane, also forms naturally just by shaking fats with water. It only takes the encapsulation of the contents of such a niche in a membrane and now you have a primitive cell. This can drift off to another suitable niche but now carrying its own self-replicating chemistry with it.

A final step is to separate the functions of genetic information into RNA's closely related chemical, DNA and the catalytic function into proteins. You now have everything needed for a living cell. It's doubtful we can ever get firm evidence for what exactly happened, but with concepts such as the RNA world, science has explained how there can be a purely natural pathway leading from primitive chemistry to cellular life.

Freya: It's still a bit of a mystery why such an insignificant chemical could organize this soup into the complex chemical systems found in life.

Max: It's only a mystery if you don't distinguish between the use of the word chemistry and chemicals. Chemistry is the process used to produce chemicals;

chemicals are the inputs and new chemicals the outputs. The unique feature about any catalyst is if you change its structure it can carry out new chemistry. This could be just doing the existing chemistry faster or slower, but there's also the possibility of a new type of chemistry able to produce entirely different sorts of chemicals. Presumably, such a leap in chemistry allowed RNA to combine amino acids into chains and make proteins. Once you have this, the door is open to produce the even more versatile and efficient enzyme catalysts.

Intelligent automatons

Freya: What you're missing from this description is how this evolutionary process selects for more and more intelligence. You can see the result of this in the way bacteria are able to make intelligent responses to their environment. They know what sort of food is present and use it efficiently to grow. When there is no food, they will shut down and become dormant.

Max: You need to be careful about how you use the word intelligent. The word 'intelligent' means the ability to weigh alternatives and work out a new solution to a problem using reason and logic. That's what we mean when we say that a person is intelligent. But with bacteria and certainly with any of the pre-biotic chemical systems, there's no creative act. Instead, they just respond according to a certain set of rules fixed by trial and error over billions of years of evolution. Even instinct doesn't really describe what they do. We say things act by instinct when they are driven by their emotions, like birds instinctively building nests for their young.

A better way to describe it is as a sort of machine intelligence. The cell is a complex but highly organized chemical manufacturing plant. Each of its components forms part of a tightly regulated network of chemical processes allowing it to reproduce and survive in a whole range of environments. It's all a matter of control over inputs, outputs and processes. You see this in the way enzymes, nature's chemical reaction vessels, are specialized so there is one for each stage in the process. They're like tiny machines able to sift through a bag of different chemicals looking for the right ones to join.

It's also well understood how the whole complex is coordinated by having <u>pathways</u> devoted to making one product. The enzyme at the entrance to a pathway acts as a gatekeeper. As the product accumulates, the gatekeeper automatically reduces the entry into the pathway and so the whole production line slows down. It's like the thermostat that shuts off the heating to keep the temperature regulated. Without any intervention, the pathway largely regulates itself and you can obtain very refined levels of control over the chemical manufacture. If you need an entirely new product, you have to switch on the genes to make a new pathway and again this is all automatic.

Freya: What's obvious is that the more complex the living system, the more layers of control and the more types of chemical information need monitoring. Only by managing this flow of information can the organism effectively respond to the challenges of surviving in ever more complex environments. This means the evolution of information processing is as important as the evolution of chemistry. Complexity of chemistry goes hand in hand with the intelligence needed to control this chemistry. This control system is itself the primordial intelligent mind we are looking for.

Max: It's just the sophistication of these controlling systems that gives the impression of an intelligent mind operating within the machinery. But of course there is no such thing, it's all built into the chemistry of the system.

Embodied chemical minds

Freya: Well if these controls and information systems are built into the very mechanisms used by the cell to reproduce and survive and aren't imposed from outside, surely what we are talking about is a sort of 'embodied mind'.

Max: I think that's really going too far. The embodied mind simply explains how the brain is necessary for us to be conscious. It's to get away from this nonsense about a mind being something other than a material substance.

Freya: It's much more than that! It's really used to describe how so many of the body's responses are built into the system without the need for conscious intervention. Things like reflex reactions, how the feet know where to go when you are walking without consciously thinking about all the muscles needing to be controlled. These mechanisms are automatic systems, but they all fall into the concept of a mind as being a computational problem solving system.

All that's happened is we have moved the concept of embodied mind down to the level of a cell. It certainly illustrates the embodied mind much better than trying to describe how the brain, with all its unknown processes, can produce a conscious mind. We've found the 'mind' of a bacteria built into its metabolism, with information passed on by chemical messengers. It's a 'chemical mind' no doubt, but it's just how our minds are built into the functioning of our brains but in this case based on chemicals rather than electrochemical systems. It's not free floating, it's hard wired into the system and in the case of bacteria, simple enough for us to understand how it can work.

Max: Well yes, the more I think about it, the term an 'embodied chemical mind' is fine, it fits in with the whole movement towards embodiment and away from mind as some strange non-material. It also strengthens the concept that consciousness only emerges from a more complex form of mind. It's just rather unusual to see it applied to a simple cell such as a bacterium rather than to the more sophisticated life forms we usually expect. OK, I'm finally on your wavelength at least with this concept and under these limited circumstances. This being the term 'embodied chemical mind' does not imply this mind is in any sense aware or conscious.

Freya: When we see evolution from this perspective, the mechanism by which enormous jumps in complexity occurs, becomes apparent. It's about the ability to control and regulate complex social networks. In many circumstances cooperation between cells with different abilities actually increase the survival of individuals in the group. The problem is how to prevent individuals from cheating, and just like any social system, you do this by regulation and control. If this high level of government adds real value to the group's survival then you can see members becoming so tightly integrated they only reproduce as a single unit. This means we have the emergence of a multi-cellular organism. So for all major stages of physical evolution there's a need for a corresponding evolution of regulation and control. The sorts of intelligence we find embodied in ourselves is the result of a further billion years of this evolutionary drive towards increasing size and complexity.

So Max, we have after all discovered a role for intelligence in evolution. Intelligence and chemistry must co-evolve if we are to explain how our incredibly complex embodied minds emerged from this primordial soup.

Chapter 3: Inherited Value

In which Freya believes she has identified how life possesses both purpose and value.

Embodied values

Freya You can only apply the terms parent and progeny to self-reproducing chemistry. With other forms of chemistry, the products bear no relationship to the chemistry making them. With auto-catalysts, the products are the self-reproducing processes themselves. New chemistry can be added to the inherited chemistry because the process of reproduction is not one hundred percent accurate. The progeny only possess this new ability. If it's of value in terms of the progeny's survival, it's going to be carried forward into further generations. If not, it's going to be eliminated from the population. This is what natural selection is all about.

With life's self-reproducing chemistry, you have the accumulated learning possessed by past generations and the potential for discovering new chemistry. In every successful life cycle leading to new offspring, there is an element of testing the value of this accumulated learning. The parents hand down the successful lessons of previous generations onto the progeny.

The point is Max, when you look at the progress of evolution you see it involves the accumulation of valuable systems and the discarding of those with less or negative value. Any new chemistry needs to add value to the whole system; otherwise, when times get tough it's de-selected. As life evolves, we are seeing the accumulation of value.

Max: Wait, hold on there, just when I was getting you back into thinking like a scientist you start heading off at a tangent into the world of values!

Freya: Why do you think it's some form of tangent?

Max: Well for a start, 'value' isn't the sort of thing that science concerns itself with.

Freya: By that, you mean physical science, but why shouldn't biology concern itself with such things? The apparent lack of any value emerging from the struggle for survival has always been a major issue. That's not only with religious thinkers but also those who use a ruthless version of 'survival of the fittest' to do what they like to the 'less fit people'. When you see each organism produced by evolution has its own intrinsic value, then each one becomes important in its own right. By confining biology to purely physical thinking, science promotes a society with no values, where anything is acceptable because only 'the fittest survive'. That's another reason for your cousins dislike of everything scientific.

Max: I still don't see why there should be any value associated with what's just physical chemistry.

Freya: Think about ordinary chemistry, say the formation of water by the reaction of oxygen and hydrogen. One bang, a release of energy and there you have a droplet of water. Now does the water contain any information about how it was made?

Max: No of course not, but I see where you think you're leading me; the genetic chemicals made by life do in fact contain all the information required to make life.

Freya: Exactly, and this information isn't a random heap of instructions, it has a very specific task of reproducing the chemistry of life in another entity. The information is of immense value to the organism. Look, you need to keep on track here as well and remember I'm not talking about a value floating in the air as some abstract concept. Like the embodied minds we find, these values are passed from generation to generation embodied into the genetic material itself. We have identified how the most basic values of life have come into the world and how they are transmitted in a simple form of chemistry. These values are all concerned with survival and reproduction of a chemical system. We derive our most basic human values from these, because without them there would be no humans to have any values, would there?

Reproductive purpose

Freya: Not only that, but we've surely found a purpose, an unstoppable chemical drive for reproduction resulting in the evolution of ever more complex chemical systems. It's a purpose requiring a chemical system to interact with the environment and the depletion of the environment forcing the selection of ever more efficient forms of chemistry.

Max: Now you've gone yet another step beyond the acceptable boundaries of science. You just can't use words like purpose and drive when describing a physical-chemical system.

Freya: But it's a purpose belonging to a naturally occurring chemical system. It's not like Darwin's 'struggle', which is only focusing on an impersonal exterior force selecting for evermore efficient and complex embodied chemical minds. The reproductive drive belongs to the system itself. It's not a universal force but a personal one. Without this interior drive, there would be no evolution. Remember evolution requires an iteration to occur between the reproductive agent and the environment, it's a two-component process .

Life's other purpose is to ensure its own survival, because without that, it can't reproduce. This is the sole purpose of its embodied mind. Its programing is all about overcoming the many obstacles the environment throws at it. This is what we all experience as a will-for-life, and it's built into the functioning of all the animals we can see, even down the tiniest of microorganism.

Max: The chemical system, however complex, is still just a machine. I admit self-reproduction belongs to this sort of chemistry, but this is still an inanimate object. It's like saying a computer program's aim is to complete the computation. But it isn't is it? It's my aim; I gave it to the computer. You can't give a purpose or a will to an unaware chemical process.

Being for-itself

Freya: The embodied intelligence developed by living things is built from the bottom up with the only purpose of ensuring they survive and reproduce. They're acting to do the work for themselves, not for others, like computers do. These computing machines are designed to be 'for-another', us in fact. They're just extra limbs and arms for our brains. That's not the case for life is it? Life is not for-another at all, it's entirely 'for-itself'.

Max: So you're supporting Dawkins and his 'selfish gene' concept? This is OK as a metaphor, but how on earth can a collection of chemicals be 'for-itself'.

Freya: For a start I don't mean it in a metaphorical sense in the way the term 'selfish' is used in the 'selfish gene', I mean life really is for-itself. Also, I don't mean to say genes are selfish, anymore than I mean a protein is selfish. Neither can we say any chemical object is for-itself. We aren't thinking about what it means to be the parts that make up a living system. If it's going to make any sense, it's got to be the whole united living system that's for-itself, not bits of it.

Max: Well I do agree it makes no sense to say a gene uses the body in order to reproduce itself, because from the body's point of view, it's using genes for the same purpose. To a geneticist, the egg is always more important than the chicken because that's what they study.

Freya: That's how highly specialized scientists can miss the big picture and where it all gets to be a bit of nonsense; dividing up living systems and saying each part is using the rest for its own ends. A living organism is much more tightly organized than that, it's reproductive purpose can only be achieved as a unified whole.

Another thought, a living system is only selfish in the sense it's for its progeny, for the reproduction of its own function in another form of itself. What's unusual is this aim can never be achieved. The offspring are never a reproduction of the parent; it's something other; even if in every respect it is an identical chemical. Life in all its complexity and diversity of forms emerged because of an embodied drive towards an unachievable goal: of an individual attempting to become its own offspring. It's an action with no achievable end!

Max: So is trying to convince any scientist a chemical reaction can have any intrinsic purpose.

Chapter 4: Living the Illusion

In which Max proves why self and free will are illusions generated by living machines.

Alternative view

Max: Well I must congratulate you on providing useful insights into the processes causing the evolution of life. You've convincingly shown how some form of chemical mind exists in even the simplest of organisms. Then you substtuted the impersonal force called the 'struggle for survival' with an interior and unstoppable chemical drive to self-replicate. Now when we have iterations between this inner chemical drive and the environment, you can fully explain how highly complex intelligent animals evolved.

Freya: So you now agree we must also have this embodied driven mind, one really struggling and caring for its own life.

Max: No, not in the slightest! In fact, everything you said just reinforces the purely mechanistic vision. There is no need for this struggling person. It is an illusion produced by a sophisticated bit of equipment. Just look at it from a scientific perspective. Mind and the reproductive chemical drive are completely embodied. This means life can go on in its mechanistic way with or without 'you'. You think you're needed, but you're not. All you think you are is just a projection of some need for self-justification. I'm afraid to say it but you're a fantasy of the embodied mind.

You have confirmed what neo-Darwinists have said all along; evolution can be fully explained by our existing science. You just built this reproductive drive into the operation of the living mechanism itself. I agree the big difference between the machines we build and these living machines is they are 'for-themselves'. Once such a self-replicating machine is let loose, it must fulfill its own reproductive programming. If, however, we built a machine with this capability, it too would be as dangerous as any living machine. So, there's no need for evolution to require the presence of an aware person at all.

Freya: How do you explain the where and why of this aware essence, this unessential self, you think we are?

Max: This is how I see it: starting with the simple chemical auto-catalyst, we wind the evolutionary process forward and get to the cellular level of complexity. We now have a self-replicating system capable of processing inputs from a wide range of chemicals with all different shapes and sizes. The living system 'decides' which ones it needs and which ones it doesn't. It can measure its own internal states and adjust its metabolism to allow it to survive long enough so it can replicate. As evolution takes its course, we see even more complexity with systems of systems overlaying and integrating structures that are ever more complex. These living systems collect and combine all the information from trillions of inputs to billions of individual cells using the central nervous system that for speed of processing has become a brain. Along the way, awareness just emerges from this brain. That's where science actually stands.

You see it's the major data processing and information handling consequences causing an aware self to emerge. But although we might wish to think otherwise, it's still entirely superfluous to the day-to-day functioning of life and therefore has no effect on the evolutionary process. It just thinks it has a function, but in the end, all the actions it believes it's making happen have already been decided by the embodied mind.

Arbitrary feelings

Freya: But you still can't answer the origins of our awareness can you? Where does, for example, the experience of sweetness come from?

Max: It's quite clear these simply are qualities the brain mechanism attaches to the inputs. The chemical food is first felt purely physically as an interaction between electronic forces at a molecular level. If the right combination of charge and shape is present, the food receptor and the food fit together and this triggers a response. But, no one in their right mind would say a cell experiences a sweet or sour taste. If cells don't need to taste for sweetness in order to make this decision, then why should we?

What's more, these qualities are quite arbitrary. For

example, certain people can see sounds or smell colors The conventional wisdom is sweetness is added on at a late stage by a highly complex brain in a rather arbitrary way. It's derived from a neural circuit which, when activated causes a suitable response. There's the actual world out there and then our mental images of it. This is the only interpretation a rational person can make based on these facts.

The main effect of this processing is to release an energetic output in the form of an action ensuring the mechanism's survival and reproduction. Your advanced brain is programmed to like the taste of something sweet, but all this happens automatically whether you're aware of it or not. They've actually shown your embodied mind makes the decisions first, and you only become aware of them afterwards. Even when you think you're making something happen by free will, you're not really acting freely at all. So you think you've decided not to eat the candy bar; really all that's happened is some other neural circuit has already been activated overriding the 'eat' command. The truth is the decision has already been made before the thought pops into your mind. It's all determined by the mechanism of your embodied mind with your awareness just tagged on at the end as some <u>ephemeral effect</u>.

Freya: You can't be right. I agree our high-level emotional reaction to a food has its origins in some initial molecular interaction. What, however, does a cell actually experience when it detects the presence of food? Admittedly, they don't perceive a sugar as being sweet, but neither do they perceive the food as a chemical, or as an interaction of electronic fields and forces. Yet it must be aware of it all the same and this initial awareness at a cellular level must be the basis of the value of sweetness we derive from this experience. This taste results from the combination of millions of these tiny cellular events after they've been processed by our advanced brains. In this way the initial event remains as the source of our experience of sweetness. This is true however remote the event initiating the whole chain of responses.

Then, what about the pleasure we get from tasting food, where does such enjoyment come from? It's in everything, visual, taste, hearing. As we've evolved into more complex organisms, the pleasure we get from these sensations has been amplified. Can't you see it's the enjoyment of these sorts of sensations by an animal that makes it want to get up and struggle, to extend its pleasure rather than just avoid pain.

I suppose in your view, when I feel pleasure or pain, my body measures the difference in electrical potential between my pain and pleasure nerves and then automatically acts to reduce the difference. Meanwhile, the feelings I have are not only arbitrary but also entirely without any function.

Max: Yes, that's just what our embodied minds do. You process information merely to increase the capacity to detect food or threats. Your emotional experience occurs after your body has already taken all the necessary steps to remove you from danger. It isn't a primary response to the threat, but a secondary reaction. I'm not saying it's without function either, since everything we do must have a reason. Perhaps it's part of the mechanism we use to lay down memories so in the future, we can avoid similar dangerous situations.

Something from nothing

Freya: All right, let's examine your mechanistic argument in a bit more detail. We agree the most basic inputs from the world come from some molecular level interactions. It could be a single photon hitting a light receptor in the eye, or a sugar molecule binding to a receptor on a cell's surface. The real world actually consists of objects exchanging energy by interacting with each other, a sort of molecular energetic buzz of action and reaction, but purely mechanical, no emotional content, no feelings.

In highly advanced life forms, this real mechanical buzzing world is overlaid by another presentation now full of emotional content and meaning. As you see it, the only problem left is to understand where within the brain's neural circuits this magical transformation occurs. There are all sorts of brain-damaged patients where certain abilities are lost. Now, by using brain scans to compare these with normal brains, you can identify the circuits generating the various sensations we experience.

Max: You see you do understand where we are in all

this. Our brains build up a picture of the world in the same way a computer constructs a picture. We build it from a series of tiny inputs captured by sensory cells, and the computer does it by processing a series of zeros and ones provided by a camera's memory card. So the pictures we make of so-called reality with all its emotional depth, feeling of involvement and beauty are all added on by our embodied minds somewhere in the final stages.

Freya: That's a good analogy because I can use it to show just how wrong you are. You see, when you took the picture, the string of zeros and ones must have been generated from something real in the first place. All the computer does is to reconstruct the image after the camera encoded it. If it's not done correctly, it ends up as a meaningless tangle. By the same argument, our emotional involvement with the world must already be in the information we're processing. It has to be there in the data otherwise the mind generates something from nothing. Just as your computer can't do that for an image, then neither can my purely computational brain generate my emotional conscious state from nothing.

Max: So that's where my analogy fails, I admit it, and as I said it's still a mystery, but the mechanism required to make this happen will be solved. The image can still be associated with the underlying mechanics of it all. Remembering my philosophy, isn't that why Descartes needed God to intervene, so that our mental images were kept in step with reality? With Darwin, we've introduced the practical necessity of survival. This ensures our images of the world and our feelings of emotional attachment to it are of practical value because it does its job to help us survive and reproduce. So we can dispense with God. Exactly how the machine generates these images is yet to be understood.

Freya: So you now admit that these images must be able to determine our actions, because if they can't, then why bother with them? I fully agree with you that without a practical use there's no evolutionary reason for these emotions. But more importantly, how can we construct such an image with its intense feeling of involvement from nothing. What's obvious is your purely mechanical world can't possibly be the reality; not even the basis of it. I agree the world we experience is an extensively derived one, but it has to be derived from something already containing some raw emotional content. If not, we are deriving something from nothing. The physicists, chemists and biochemists are abstracting and dissecting this actual world and presenting it in terms of quarks, atoms, enzymes and cells. OK, I admit these concepts are useful, but don't fool me into thinking they're seeing reality either. This, what we both touch, taste and feel is more actual than their abstracted dead world.

Facing reality

Max: So that's your version of how *you* got to be here, in the world of emotional experiences and involvement? You're claiming that what scientists describe as forms of energy, taking on the shape of various physical entities such as sugar molecules and light, are experienced by life as an emotional feeling.

Freya: Yes, I guess so. This isn't just a brain thing at all; it's present at all levels of life. Even cells are feeling the presence of the world and responding to it, it's just they perceive it in a very limited way. What advanced life does is to evaluate and concentrate all the feelings from billions of little energy inputs received from the environment using all sorts of sense receptors. The more advanced the life form the more range of inputs it has to work on, the more potent the experience and the more vivid the reality. Your mechanism is just a description of the process devoid of the emotional component. That's why it can't describe what it's like to be real and *in* the world. That's why I become some strange useless phantasm floating like a disembodied ghost over the body.

Max: So your alternative is to have emotional awareness grounded in our molecular interactions with the world. You've pushed the origins of our experience down to the input of raw forms of energy. The problem is your solution doesn't solve the issue of being aware of the world at all. All you have done is to pass it down to this lowest level. This is good because now you can see how absurd your whole notion is.

Freya: Why do you say that?

Max: Because you still need life to transform energy into feeling. Whereas I had this happening at the highest levels of mental processing you now have it at the lowest. But this means you're still saying life is somehow different than the rest of the physical universe. The problem is there's no distinction between <u>living chemistry</u> and ordinary chemistry. If life is to experience exchanges of energy emotionally, then it must also be true for any chemical as it interacts with energetic inputs. We don't stop there either, because it must be so for every physical process, electrons, protons, quarks even. These are in a perpetual state of energy exchange. For your impassioned argument to be true, you need the entire universe to be full of things with passions and feelings. So now the energy of physics is to be replaced with emotional feelings and instead of energy $= mc^2$ you have emotion $= mc^2$.

You need to face reality. You've solved the problem of how our aware self evolved OK, but in order to do so, you have just overthrown the whole of physics. You've taken us back to the middle-ages with its universe composed of spirits and fairies. Your line of reasoning will have us return to the debate about how many angels fit on the head of a pin!

Part II: The Dialogue

Chapter 5: Origins of Experience

Freya discusses the problems caused by her search for self with Orin, a philosopher-scientist and they discover <u>consciousness has its origins</u> in events raised when simple chemicals process inputs of energy.

Dangerous nonsense

Orin: So, let me get this into my head. Max thinks he killed your whole argument by showing living things can only have real subjective experiences if non-living things also have some form of experience. That is, even inanimate objects would need to be aware in some way. His argument is science has convincingly shown life is a form of chemistry and so no different from ordinary chemistry. The only difference is life is just a more organized and complex form of chemistry. Since the inanimate basis of life is not aware then neither can a primitive life form be aware.

Freya: The argument used by the mainstream evolutionary theory is our consciousness emerges from the complexity of this chemistry. In evolutionary terms, awareness and consciousness are late stage add-ons.

Orin: The next movement in this argument is to say that because consciousness plays no part in this underlying biochemical machine, it isn't a necessary component of the body's function. This seems to be confirmed by experiments showing how the body makes decisions even before we are aware of them. This leads us to conclude our conscious self is a secondary effect: in the worst case our self-consciousness is an illusion produced by this biochemical machine we call life. This illusion cannot affect the processes of the machine it just thinks it can. The machine is fundamental, its mechanisms determine all the decisions we think we are consciously making. This means even our free will is an illusion.

Freya: The logical end to his line of reasoning is unspeakable nonsense. It says the only thing I can be sure of, my own existence, is the one thing that's not real. He knows it's not true; we all know it's not true. I know I really experience the world. I feel I am here, in it good and proper whether I like it or not. I also know I can change the world by what I do and within limits, choose to do what I like. Yet his logic is sound, it just ends in a ridiculous conclusion. How can that be so?

Orin: It is also <u>dangerous nonsense</u> because once you conclude your free will is an illusion you open the gates to all sorts of destructive actions because no one is to blame for what they do. In any given situation, we simply do what we are programmed to do by the survival machine operating us. That survival machine just processes the physical inputs it is confronted with according to pre-programmed routes, so the outcome is determined, even if the actual programming is complex. This means our whole concept of justice based on a person being able to act with free will is wrong. When this is under attack then so is the basis of law and order underpinning our entire civilization.

This argument is even more dangerous because I really know I am real and free to act, but here I have the scientific authority telling me other people are just machines. It is an open invitation to express every psychopathic tendency we can imagine. Science provides an unscrupulous leader with the argument "although members of my party, tribe, race, nation, religion etc. aren't machines, the others obviously are, therefore we can treat them as we like".

Freya: I want my consciousness, my self and my free will to be restored to reality and if this means giving consciousness of some form to all animals, then that's fine as well. But I've got to say even I can't quite bring myself to make a leap into saying that inanimate objects are also aware.

Orin: Well if this means you need to go beyond the living for the origins of your consciousness and descend into the inanimate, into the furthest realms of physics, then so be it. If it is so, then we will be describing a sort of Copernican revolution applied to all of science, with our position as the only entity having meaningful awareness put into a proper perspective. Our awareness now becomes an outgrowth of something pre-existing and much, much deeper!

Freya: And this is where Max will say that I'm heading off into the land of spirits and fairies instead of atoms, quarks and strings.

Agreed

Freya: I feel a lot of problems are caused by the rather narrow subject matter we scientists have to specialize in. You know, the case of scientists knowing more and more about less and less until they know absolutely everything about nothing. Given your broad experience of philosophy, I was hoping you would be able to help me through some of the philosophical issues raised by our discussion.

Orin: Well I am sure that Max will argue that as a philosopher I know less and less about more and more until I know <u>absolutely nothing about everything</u>! So perhaps between the two of us we might actually get to know something about conscious life and its origins!

The best place to start is simply recalling all the points you and Max did agree about. Then we can explore the consequences of this great rift you have discovered at the foundation of physical and life sciences.

Freya: This is my take on it. We're both happy to accept the current view that at the base of the evolutionary drive there is an auto-catalytic mechanism, the self-replicator.

Orin: I think Max had a great insight when he saw the process of mutation and selection results in an increase in diversity of the chemistry available to a living system. With inanimate things, the chemistry making them is not preserved and instead chemical products just accumulate. With life, the chemistry is both preserved and replicated. I mean by this the enzymatic reactions, metabolic pathways and cellular processes. These chemical processes are passed on to the living progeny as they replicate themselves.

Freya: Also to be a self-replicator, you need to preserve the organization of the system. This means in living systems, the chemistry itself needs to be careful. Careful of what it takes into the process, careful about how it processes the product, careful about how it reproduces itself. Such a concept is not at all applicable to non-living chemistry based only on the chemical properties of the inputs. This form of chemistry does its job but makes no attempt to control its inputs and outputs so usually produces a disorganized soup of chemicals.

Then we can all agree life is organized into a hierarchy of biochemical and cellular sub-systems. The operation of the lowest functional living unit, the cell, is underpinned by a complex and highly regulated network of biochemical and molecular biological interactions. To get multi-cellular life forms such as us, you find a further organization of cells and division of labor into organs. At all levels, controls exist to coordinate all these biochemical and cellular sub-systems. This could be via chemicals acting as messengers between cells, or the electrochemical waves in nerves and neurons connecting distant parts together.

Life has evolved from the ground up, from the inanimate molecular level upwards in terms of bodily size and complexity. New abilities emerge as the complexity of the organism increases. This is the concept of <u>emergent evolution</u>. We're the result of nearly four billion years of biochemical process development of a system designed to survive and reproduce in the widest number of environments.

Orin: I thought it was good how you both saw evolution as the result of iterations of auto-catalytic systems with the environment. Iterations are not cycles, there is no equals sign and so the process is a development of mutually enforced adaptations. You also realized how the concept of a 'selfish gene' was the result of highly confused thinking. That breaking up the whole system into bits, then arguing the bits are more important than the whole, is such a basic mistake to make, but one that scientists do all the time.

Finally, you addressed the question of how such leaps of complexity could occur, and to do this you used co-operation between a living organism and its close genetic relatives. By forming an increasingly mutually dependent and interlocked social system, each member can perform its own specialized task. If the entire system is locked into a single replicating unit, a new integrated self-replicating super-system emerges

Freya: Colonies of insects where only the queen replicates are examples of a system mid-way between being a social group of individuals or a single organism. Without the involvement of cooperation, it's hard to explain how competition alone could cause large scale increases in organizational complexity. So selfishness and collaboration are all built into the process of evolution.

Orin: In your search for the origins of intelligence, Max did reluctantly agree to the concept of there being an embodied chemical mind in even the simplest of life forms. Higher life forms with a brain would use this as a basis for their embodied electrochemical brains. To him, this embodied chemical mind was similar to that of a machine or computer in not being aware. Awareness arises from a highly developed form of embodied mind, presumably at the level of an advanced electrochemical brain. It is because of this he sees life as basically a robotic mechanism. Any awareness and consciousness emerge at later stages in evolution and therefore it is not essential to that process. How this awareness arises from such a mechanism is not understood, but given enough time, it will be.

Your solution is to have some form of awareness in even the simplest of embodied chemical mind. This raises a real problem, where do you stop in this downward track. You cannot stop at plants, microbes, or even the chemical origins of life. You need to go all the way down or else awareness still becomes an add-on. You end up saying the primary stuff of the universe contains an experiencing element. This is the only way you can explain our intensely emotional experience of the world. That is a radical thought.

What are we doing?

Freya: No wonder Max had problems, just as I and

Stephen J. Brewer

any other reasonable person would.

Orin: Being reasonable is what we all like to be; after all, we are trained to believe reason and logic are the ultimate achievement of the rational mind. In the end, however, you should always follow your intuitions about these sorts of things. Because if you look closer you will find these are themselves the source of all reason and scientific advancement. These feelings are the output of your embodied mind with its inbuilt intelligence resulting from billions of years of evolutionary development. Our emotions, feelings and intuitions are all reasonable; they have to be, otherwise we would not have survived.

You feel something is not right with our <u>theories</u> about the origins of self because as it stands, it does not make a consistent picture about your actual experience of the world. I also feel the same about it; all things need to be consistent with each other. This faith underpins all scientific and western philosophical reasoning. There is a reason for everything and we can if we dig hard enough and see things in the right light, find those reasons.

Freya: But why should things need to be consistent? Perhaps it's fine to have things pop out of nowhere. The universe apparently did just this according to the 'big bang' theory. Perhaps it's impossible to explain everything; maybe we, like the universe, don't have our origins in any pre-existing state, we could be something entirely new, an unprecedented phenomenon. Perhaps life, mind, consciousness can all just pop into existence as complexity increases.

Orin: This is the concept of <u>strong emergence</u>. The problem is it actually introduces a whole series of unprecedented states with no discernible cause in the previous conditions. Now to my mind this is simply the same as saying 'something magic happens' and for example, matter becomes alive, and then somehow life becomes conscious. This is actually counter to the demand for science to provide an entirely naturalistic explanation. We can only do this if everything is linked to everything else by some form of causal relationship. What is more, not only does everything need a cause, but it must have an impact on others. We, as aware subjects are not only caused by events in the universe but we must also have an effect on it. If either of these is not true, then we are indeed in some different place than physical reality. If we successfully establish this relationship and show how it works, we will be able to better describe the reality of being. It is about improving the explanatory power of our scientific reasoning so it can happily include not just the physical nature of our being, but also our psychical being, our self-consciousness.

Freya: All the same, perhaps science's faith in reason is unfounded and we are the proof of it?

Orin: It may well be so, but we should at least try. The big bang causes a lot of soul searching amongst physicists who are trying to explain its origins in terms of higher dimensions. This means it is even more important we attempt to explain such a glaringly obvious rift between science's description of the world and the one we experience. What is more, this rift drives a wedge between science and the rest of humanity. Not only does it seem to deny our reality, but it also denies free will, the very basis of what makes us human.

In the end, your call that the self should have a role in the evolution and functioning of your body is based on an intuition there is something wrong with the way science sees the world from a purely physical and objective perspective. The fact is I am here, along with a whole host of animals and we obviously experience the world in a very real way. If it means for this to be true, everything, animate and inanimate must be aware, then so be it. Perhaps our scientific worldview needs to be changed.

Not self-aware

Freya: That's all very well of course, but how do we get ourselves to understand what we mean by saying 'everything is aware'.

Orin: I think we start out by asserting that however we do this, we must not invent some new physics. We could add a new perspective to our existing physics, but we must not invent a non-physical force or essence. This will just land us back with the same problem. After all, it is precisely the concept of consciousness being outside the physical description of the world we are trying to overcome.

Freya: We have to see how the primordial origins of

what we call consciousness, awareness, experiences and emotions can be found in inanimate entities obeying physical laws. How do we do this?

Orin: I suggest we start at the top with our experience of consciousness and then work our way down to see if we can get to some conception of the sort of awareness lying at the route of our own conscious states. It has to be something universally present in everything.

Freya: The next obvious point is we aren't on about these things being self-aware. Self-consciousness seems to be possessed by very few creatures. The only other beings we know to be self-aware are other humans; although there are other animals that seem to pass the test.

Orin: What test is that then?

Freya: Without the animal knowing, you dab some rouge on its face then show it a mirror. If it sees the image in the mirror as itself, it will attempt to remove or touch the rouge. If they do, then they have passed the self-consciousness test. Humans aren't born with this ability and it only appears when we are about 18 months old. Some apes can pass the test as well, but most of the time an animal will have no knowledge the image is of itself. This means most animals aren't self-conscious, although of course they're still aware of things going on around them.

Orin: You can see how self-consciousness results in this peculiar feeling of being outside your own body. After all, you have this strange ability to see yourself in an image outside your body and this could well be the origin of the belief that the body and mind are separate. Now, I think when scientists say consciousness is an illusion, they may well be referring to this self-conscious state, and in some aspects this may be so.

The other problem is we tend to imagine consciousness is about thinking and reasoning. I do not see how you can do this without some very advanced form of language and again, you only see such language with self-conscious humans. Even then, we do not spend a lot of our time thinking or reasoning about what to do. Mostly we just act in response to our emotions. Afterwards, it is true we might apply reason to explain or justify our actions.

Freya: We don't apply reason to seeing, hearing or tasting. These feelings make us aware of something going on requiring us to act in some way. I might say, 'I'm hungry because I haven't had any food since breakfast'. But I don't need these words or this reasoning process in order to experience hunger or to act to stop that feeling.

Orin: The first and possibly the most difficult step is to get away from all the special things self-consciousness allows us to do. We need to see these as additional capabilities added to our underlying consciousness. Without this underlying awareness there would be no self-consciousness.

Freya: You can see many animals not considered self-conscious are obviously enjoying consciousness, sunning on the rocks, flying, hunting, preening and swimming. Look at human child under the age of 18 months and remember although it's not self-conscious it most certainly is conscious and quite able to let you know whether it's feeling happy or sad, what it wants and when it wants it. Even plants are conscious of the sun. They will follow its path by bending its leaves towards the sun. They're also conscious of water and nutrients and their roots will follow these into the ground.

Orin: I agree all these animals are meaningfully present in the world, just like us. Or perhaps even more so because they know where they are without loosing themselves in reason or existing outside themselves as some image in a mirror.

Conscious of what?

Freya: And the next point surely is these entities need to be aware of something. Some event must happen to them if they are to be aware of anything.

Orin: In a universe where there were no events, there would be nothing of which to be conscious. Perhaps it might be argued a disembodied consciousness could be aware of nothing, but we have already rejected the concept of this type of awareness. So we must argue from our evidence for an embodied mind that for it to be aware, something needs to stimulate it. If not, it will just remain dormant waiting for an input to process. Of course, with our advanced embodied minds, things are occurring all the time. We are in a constant state of agitation and adjustment, flooded with events and stimulations, with only a few of them coming into full consciousness. Even when sleeping, we are still continually monitoring our environment and ready to wake if something unexpected happens. If there were no external stimulations, something will be going on inside. This means our so-called subconscious processes are continuously working for us by being aware of our internal states and responding to them in a meaningful way.

Freya: If we are to argue inanimate things are aware, we can safely argue the cells making up our bodies are also aware. What they are aware of, however, will depend on the role they play in supporting the overall living system. This comes down to the specialist equipment a cell possesses. All cells must have their own general biochemical system keeping them alive and functioning. But if they are to play a part in the greater organization, there will need to be a way in which they add value by having an additional specialist function. This function will need to be activated by a specific input from its environment.

The range of things an inanimate entity would be aware of would be limited to the sorts of inputs it can process. And given the simplicity of their structure, these are going to be of a very limited scope.

My awareness

Freya: The next point is that my awareness is a personal thing. It belongs to me and me only and what I'm aware of is focused on those inputs required for my survival and reproduction.

Orin: You described this sort of system as being a personal one. This means we should concentrate on finding a basic awareness in inanimate systems that are also individual. That is, the system has to be organized as a single united entity. We are not expecting to find any sign of awareness in piles of rubble or aggregates. This is because the components do not interact with each other in a systematic way.

We also do not mean weather, planetary or ecological systems are aware. These are held together by their external relationships it is true, but that makes them similar to a social system. We should concentrate on systems <u>totally united</u> by their internal relationships. We also are not looking for some dispersed general form of awareness not attached to anything in particular. All this does is to return us to a mysterious entity defying any physical laws. The awareness we seek will be embodied like ours. Any experience it has will be the result of some process carried out by its own unified system and experienced only by that system.

There is also no reason to think this awareness has any direction or aim to it. Inanimate systems do not aim for their own reproduction in the same way living systems do, they are not for-themselves as we are, or for-another as machines are for our use. It will be a purposeless form of awareness, but this can still make it the basis of our own purposeful awareness.

I think this means we confine our attention to the individual atoms, chemicals or molecules. These are the simplest fully integrated components we can isolate from the disorganized aggregates making up our physical world. They are held in unity by sharing electrons forming into integrated clouds joining all parts of the structure. It is because this unity of electrons can only be formed under the strict rules described by quantum mechanics, elements such as carbon, hydrogen and water, can combine into organic molecules. Only when all the electronic needs are satisfied is the resulting structure stable. In the world of chemicals, there are a vast number of highly structured molecules meeting our criteria.

Energetic events

Orin: This electronic cloud can interact with other electrons or electromagnetic energy such as light, and in so doing, it will first absorb the energy and after some form of processing, it will re-emit it. The simplest chemical systems can only respond to a very limited range of inputs. Perhaps a hydrogen atom is a good example. When it absorbs the energy in a photon of light the electrons jump to a higher energy state. Because this is unstable, the electrons drop back down to the lower energy state when the energy is released as a photon of light. This jump can only happen if the light has just the right energy, so it meets the need for its response being determined by a specific input.

Freya: So, we are saying the basis of our awareness is simply down to our need to chemically process energy inputted from the environment.

Orin: Exactly, we are built from a highly organized system of molecular scale energy-processors. These are continually responding to inputs of energy and processing these events. Chemicals do not respond to every form of energy either, but only certain types and how they respond and what they respond to depends on their own structure. This means molecules are able to discriminate between different forms of energy. Each 'event' will produce a pulse of energy with its own unique signature. Also, since it belongs to the system responding to the event, so it is a private event. These event-based pulses of energy have to be the source of our awareness.

Freya: By making our awareness dependent on molecular systems processing tiny inputs of energy, we can draw a link between our super-advanced awareness of the world and the actual occurrence of such events in even the simplest inanimate chemical systems. This also means we can only be aware of the world through the <u>experiences</u> caused by such tiny molecular level events.

Orin: The great advantage about this is all the physics remains untouched, all the quantum mechanical descriptions remain in the same place, the laws of thermodynamics and gravity are not affected in any way. All we have done is to say these energetic events are the basis of all experience. Physical science naturally follows the physical component and the experiential component is not at all important. However, the experiential component of these energetic events cannot be ignored if we are to fully understand the life sciences. That is because what was trivial for the physical sciences is now the key component of the embodied mind that evolved to take living systems into such high states of consciousness.

Where next?

Freya: This is the hypothesis. How do we test it?

Orin: We have not changed the physics, so it is not testable by scientific experiment. The only argument for accepting it is that by adding the experiential component to events we get a better description of what we find in the world. Rather than calling it a hypothesis, you should see it as a postulate, a basis we can use to construct a picture of the world. In the end, all our theories are based on a number of postulates. In mathematics these are called axioms, in physics these are called principles. Einstein's theory is based on the counter-intuitive principle that however fast you are traveling, light travels at the same speed. Euclidean geometry incorporates the axiom that parallel lines never intersect. They are assumptions we cannot test nor prove within the theory itself. What they do is allow us to construct theories by which we can grasp and understand the world.

This postulate, that a pulse of energy is equivalent to an experience allows our science to explain our real presence in the world. It does this because rather than seeing life as a purely physical processing of chemicals, we now also see it as handling and organizing experiential events arising from energy inputs. Living embodied minds are evaluating and concentrating these bits of energy until we experience them as a powerful emotional state. This emotion is still a package of energy and it must be released in some way or another.

Freya: Life is no longer just a highly complex chemical factory but it's an organized processor of tiny energetic experiences finally outputted as actions aimed at survival and reproduction. These are felt as emotions; literally these become our motivating forces. This means we're even describing the origins of psychology itself.

Orin: That is right, and we need to integrate this primordial psychology into the evolution of all life forms, right back to the beginning, not just at the advanced human level. If we are to really understand how our consciousness is embedded in the world it is essential to see life as a system for combining and amplifying experiences coming from trillions of these tiny throbs of energy.

Freya: So, Max's nonsensical and very dangerous conclusion I'm merely an illusion, is replaced by me being the ground of all reality. The real illusion is then his purely physical interpretation of the world.

Orin: Beware! You've just fallen into the trap of saying only your subjective experience is meaningful. This is the track taken by the idealists. This occurs when you start from the premise that without the conscious mind there is nothing to think about. This is the error Descartes' made right back in the 17th century when he concluded 'I think therefore I am'. He started the line of reasoning that eventually ignored the physical side of our minds and only pursued the awareness component. This means that in total opposition to Max, you conclude all physical things are an illusion generated by our minds. The next step is to conclude that everything is just something dreampt up by the mind of God so nothing you do in this world is of any consequence.

Freya: I see, when you follow just the physical component you get to our conscious minds being an illusion, and when you follow the purely experiential track, you end up with matter being the illusion.

Orin: We have to avoid following just one of these routes because both the physical and the experiential side are essential components of our embodied conscious minds.

Freya: Now, what you're saying makes sense. By adopting this principle, we have to end up with a theory fitting the reality of the world as we experience it. That includes both its physical component independent of me, and the experiencing component making up my inner self. Then we need to see if the continual flow of action and reaction between these related states can more accurately describe the reality I actually find myself living in.

Chapter 6: Enjoying My World

In which Orin and Freya explore how an organism's channeling of energy for survival and reproduction reveals a world of pleasure and beauty.

Motivational forces

Orin: In our search for the origins of our awareness of the world, we need to consider how we derive subjective feelings and emotions from simple chemical events. From a physical perspective, there is no difference whether I input energy into a stone by tossing it into the air, or put energy into an animal by giving it food. In both cases, the energy inputted must be acted upon in some way. If the stone bounces after hitting the ground, some of the energy of motion will be turned into heat but the rest will be used to continue its trajectory. When an animal feeds, any energy not dissipated as heat must be outputted as some form of motion. The difference is life channels this motion towards its survival and reproduction. This movement might be to just direct the flow of electrons to make a biochemical, but the one concerning us is how it produces a physical movement of the whole organism towards food or away from a predator.

Freya: We experience this <u>motivational force</u> as an emotion, literally the feeling we get that's so powerful it must be released in some form of physical action. The form of our emotional feelings are then a highly structured response based on the 'wiring' of the embodied mind after it's analyzed and channeled the energy received from the environment. The animal's outputs are structured by billions of years of natural selection. This makes it do the right thing in a given set of circumstances. Therefore, emotions aren't at all irrational, but highly focused outputs of energy formed by the embodied mind demanding we act in a specific way.

Orin: These outputs do not require any rational thoughts. No language or reasoning with concepts is required. Such rational thinking may distinguish us from most of the animals we are dealing with but this emotional awareness is nothing to do with thinking at all.

Experiencing energy

Orin: We need to outline how life focuses these tiny energetic events at a molecular level into a powerful emotional output. This we must do without inventing any new physics, biochemistry or biology.

Going back to basics, when two systems interact energy must be exchanged and this results in two events being raised, one for each part in this exchange. Of course, how both systems respond to this exchange is entirely determined by their internal constitution.

Freya: Thinking of it from the perspective of a simple living cell, if the input is its food, say a sugar, the first event is raised when the cell's sugar receptor binds a sugar molecule. It's just like a trap, the sugar receptor looks like a good place to be if you're a sugar molecule and it fits into the space provided by the cell like a hand in a glove. As soon as it settles in, the trap is sprung and the sugar is whisked inside the cell and quickly dissembled into its component bits. During this stage, the cell will experience a whole string of biochemical events spreading throughout the entire cellular system. This external entity has set in motion a whole series of internal events.

Once we recognize this, it's easy to see how multi-celled organisms generate our more powerful motivational feelings. A single cell constructs its physical feelings towards its own survival and reproduction. In a multi-cellular organism, a cell directs its actions towards the survival and reproduction of the entire organism. To achieve this, many cells must cooperate. You can see this occurring by a mechanism called quorum sensing. This happens when a group of identical cells are stimulated to release the same chemical messenger. When enough cells are activated, the entire community of cells start to coordinate their responses. This can be waves of contractions causing the organism to move towards the sugary food's source. This is just an example of how tiny chemical inputs can be transformed into information, which is then channeled into a physical response aimed at reproduction and survival.

When you get to organisms that are even more complex, the events raised by such cellular networks may need further evaluation. This usually involves raising powerful electrical signals traveling along nerves converging on a central processing region we recognize as a brain. Since these signals arrive from all over the organism, this allows information from other cellular networks to be compared, amplified or suppressed. Any outputted signal is an even more powerful motivational feeling on a scale needed to channel the organism's energy into some specific form of action.

Information processing

Freya: The problem is we seem to be describing a purely mechanical system. Our emotions are attached to these inputs of energy in the same way the wheels of a car are attached to an engine.

Orin: That is not really so because you have the input's energy providing both physical force and information. Any energy will need to be transmitted from the environment and as such, it will have both magnitude and direction. You can see it like an arrow, and that is just how physicists represent it; an arrow of a certain length equal to the energy's magnitude and pointing in the direction it is going. The important factor here is this energy also inputs information about the world. When we see this, there is no problem in how the embodied mind can both compute and focus energy into desirable actions.

Freya: I've heard physicists say information and energy are equivalent. The problem is I'm never quite sure what they mean by this.

Orin: It came about by needing to support the second law of thermodynamics from a thought experiment produced by the famous physicist, James Clerk Maxwell. This law is about how hot things become cold, but never the reverse, so lukewarm water does not spontaneously separate into hot and cold components. It is essential to our entire understanding of how energy drives the physical world. Maxwell imagined a 'demon', an entity able to measure how fast molecules were moving and by acting on this information it separates the fast moving hot molecules from the slow moving cold molecules. This means we could use information to contravene this cornerstone of physics. The solution is simple; to recognize information is itself a form of energy. When you now add up the information content and the heat energy of the system, everything balances out. The second law of thermodynamics is saved and physicists can rest easy in their beds.

For us this is important because chemistry is about how chemicals process discrete packages of energy. This means we can also say it is about how chemicals process information. All chemicals are in effect information processors. Since these energetic events go on all the time, we can also see the universe as composed of a network of information processors. It is just there is no organization to this network, no goal to survive and reproduce, so the information-energy just gets dissipated as heat.

Inputted energy is also the data used by life. Whether it is light, sound or smell these are vectors having both size and direction. Life can use the direction to determine where a threat or opportunity arises and based on a purely physical response to the forces involved, move accordingly. If, however, we focus on the size of the event to extract what are known as its scalar quantities, the energy looses its compelling force. We can now combine data from different sources by adding, subtracting, contrasting and transforming it without being forced into actions. In effect, we can evaluate these forces without needing to act on them.

This evaluated information is still a form of energy with a particular emotional content. As such it can be transformed into a vector channeling this emotional energy into a specific action. The equivalence of energy and information means we can change action into thought and thought into action. Transformation of vector into scalar quantities and reversion into vectors ensures our actions are not simple mechanical outputs but always directed towards achieving our own ends.

Freya: I suppose you're saying like most modern cars, there's a computer between the wheels and the engine. Unlike a stone, our embodied minds allow us to actively resist or exploit the forces the environment throws at us.

Correct actions

Freya: I guess the problem with processing information separated from the physical context is it can introduce errors. For example, when cells in our taste buds detect sugar, a patterned electrochemical event assigning the food a high value is sent straight to the brain. This elicits the whole cascade of events required for us to consume the food. However, the firing of these taste receptors is only correlated with the food's actual energetic value. This means we can fool our bodies into eating useless low calorie sweeteners masquerading as high-energy sugars. Of course, even though the sweetener tricks the taste buds, it doesn't trick the cell. It's not going to produce the same physical response as a real sugar because it won't produce the correct responses.

Our cellular based chemical minds are not so easy to fool and fortunately, these will also have their say. So, if I tried to live on the low calorie sweetener, my taste buds might inform me 'high energy food eat it' but my muscles still tells me 'fatigue' and my stomach signals, 'hungry'. These tiny cellular feelings when used to recruit millions of others output powerful emotions forcing us to respond.

I suppose, what's important is the feelings telling us about our cellular state don't contradict the feelings from the information processing route. Perhaps we can see an important role for the higher levels of information processing we find in more complex organisms. By comparing the size of the feelings derived from various sources, we can check for consistency. My advanced mind then acts as a high-end error checking process.

Orin: Although using derivatives of data makes these information-processing systems prone to error, the data is still from the same world. In this way, our minds are not disconnected from the physical inputs.

For inanimate entities, there is no correct action, just an outputted action based on the physical processing of energetic inputs. For living organisms, there are correct actions because these must maintain the living system. This requires the information content of the input to be processed. The ultimate check on the relevance of its information processing is natural selection. The intelligence and logic of the processing must ensure these valuations are not only 'calls to action' but produce the correct action in the given circumstances.

Now, Max might say since the transmitted signals are of an entirely different type to the input, we have introduced a whole series of discontinuities. The original experience is then totally lost and the final output of emotional-energy is still disconnected from reality. My response is 'not so', because you can accurately pass on information about the current state of a system using any sort of code. Whether you use chemical or electrical waves as an intermediary, it makes no difference. The signal can be recovered intact using the reverse of the signal encoding system. Sound goes to electricity, which goes to light, which goes to pits in a disc and you have a CD. The CD player just reverses the flow to end up with sound-energy patterns just as the original, even if distorted to some extent. In these cases, a connection occurs between both parts of the system regardless of the medium used for sending it.

Freya: We are arguing our emotions are about how it feels to have channeled pulses of energy building up in our bodies. The channeling occurs because our embodied mind can extract the purely informational component of the inputted energy-data to reform the energy into an action. If we don't need to take any action, then the energy is released as heat. The bigger the organism, the more conditions it needs to take into account and the more powerful and complex the emotional output will be.

Orin: This call to action is what makes us in-the-world, it is how our embodied minds make us respond in a focused way to events occurring in the real world. We humans may like to claim we are in control of our emotions, whether this is so remains to be seen; but for the animals we are talking about, they are entirely controlled by their own emotions. The basic way any animal responds is determined by its overriding goals of survival and reproduction. It is still self-determined, a response directed towards their personal well-being and the outputted action they cause may work with or against the prevailing external conditions.

Freya: But if all our actions are determined with this goal in mind, we aren't free.

Orin: For the moment, it is important to focus on the nature of a purely emotionally determined response because it is through these we can unite our feelings with the world. We should put on hold any discussions about how certain advanced animals are able to override these emotions using reason. The truth is the majority of animals have their actions directed by their emotions, not by reason. We share with the animals experiences of pain and pleasure, it is just they do not ask why. They feel a certain way and act accordingly.

Freya: I think we've outlined our current understanding of what physically and mentally goes on to convert the vast numbers of simple chemical inputs into such complex emotional responses compelling both humans and animals to act in certain ways.

Revealing quality

Freya: What puzzles me, is why these events feel the way they do. It appears we somehow and rather arbitrarily assign a quality of sweetness to the feeling generated by tasting a sugar. In fact, because these are subjective, there's no way of even telling if two individuals actually have the same experience. We both taste sugar, but is your experience of sweetness the same as mine? Perhaps you taste it as bitter. I suppose it wouldn't really matter, just so long as we consistently attach the same feeling to the same event. If not, the result would indeed be chaotic. Also, from a practical point of view, because all life is genetically related, it's highly likely we have information processing systems yielding very similar experiences from identical inputs.

Orin: This gets us to the old philosophical problem of where the 'redness of red' and the 'sweetness of sugar' come from. I am not at all sure it is something we can solve here, but one thing we can state is that these qualities do not belong to an object. Instead, they describe experiences we derive from the extensive processing of light or chemical energy arising from say, seeing and tasting a red apple. The specific experience of the color red arises from the processing of the pattern of energy associated with light of a certain wavelength. In the same way, sweetness arises from the processing of electrical signals derived from the recognition of sugar molecules by taste-bud receptors. These properties are not entirely in the input, red light is not red, and sugar is not sweet in itself but neither is it entirely something we have made up from scratch.

You argued with Max why the world we experience, full of color, beauty, sound and taste, has to be a revelation of an actual state of the world. For this to be true, these qualities must already exist for us to discover. Max maintains these qualities are entirely inventions of the mind. The problem is his interpretation requires our minds to construct something from nothing.

Now your concept of discovering something preexisting also has difficulties. In the same way America was not invented by Columbus but discovered, we must also discover these qualities. If this is so then where are they held? You cannot have a 'look-up' table in your mind attaching these special experiences to the inputs because these would still need to be derived from somewhere. My solution is to have these qualities emerge from the data processing itself. It sounds strange but fortunately, you can see a precedent for this in mathematics. You see, new mathematical structures are only found by applying strict laws of proof and logic. This means mathematicians are not considered to have invented anything but to have made a discovery. For example, the next highest prime number might not be known, but it can be discovered by testing every subsequent number to see if it is divisible by any number other than one. In the same way primes await discovery, qualities would be discovered as the specific experiences associated with particular structural forms of energy-data. The structures would be generated based on the original form of the inputs and the subsequent processing by the embodied mind.

The important point here is if we allow qualities to be invented, we contravene the causal principle; every event is caused by some previous event. Neither we, nor for that matter science, can allow anything to appear from nowhere. Instead, any entity must reproducibly emerge from the processing of what it is given.

Freya: I think I can just about see where you are coming from but what's really important is the quality of experience which helps us survive. We have information processing systems such as pain receptors whose entire function is to stop me doing stupid things. Then I've another group of receptors and pathways rewarding me for taking actions increasing my chances of surviving and of course reproducing. The qualities of these sensations are an essential part of my being able to survive.

Drawn in

Freya: I suppose when my body is in total harmony, it produces no emotional states because no actions are required. This doesn't mean I have no awareness of the world, since events are still processed for their informational value. It's just I don't need to output any actions. Perhaps when all desires are satisfied and we only have information processing, all animals, not just us, have a purely aesthetic appreciation of the world. Why this informational display is beautiful has always intrigued me. I again believe it has practical value because it draws us into the world to seek ever more intense experiences. After all, survival is active and requires movement towards this end. This means seeking pleasure and experiencing beauty are the great drivers for our action on the world. If we're not drawn into the world; if we just sit back and let it come to us, we would go nowhere and be out-competed by those animals positively seeking experiences.

I think this demonstrates how the real cause of the struggle at the feeding trough is a subjective enjoyment of the experience of eating itself. Because of this drive, evolution has made us into collectors of enjoyable experiences. By spreading a wide sensory net, we gather pleasurable experiences and act to maintain ourselves within that realm. This means the real driving force behind evolution is the intensification of our subjective enjoyment of the world.

Orin: We may have our feet in a world about which we can do nothing, but we can process this raw data into a form bringing us intense pleasure. The qualities uncovered by our embodied minds turn an indifferent world of physical forces into our enjoyable world full of rich experiences.

Chapter 7: One of Many

In which Freya discovers the wholeness of her being and how she is killed when dissected by science.

Limited science

Orin: Physical science can only study the reproducible properties of objects. Your <u>subjective state</u> is not one of these and so falls outside its realms. This is fine until scientists try in vain to make a connection between the physical structure called a brain and consciousness. The other problem is they continue to treat living organisms as if they were inanimate objects and study them by chopping them up into bits. This makes it impossible to ever re-discover the animating living subject. A moment's thought, and this would be obvious, but they propagate the ridiculous conclusion a person is either unnecessary or the illusion of a machine. What makes it even more ridiculous is the scientists claiming you are an illusion must accept they too are illusions. Therefore, the world they study must also be an illusion generated by this illusionary observing subject. So why should we believe any of their claims to discover the truth about the world?

To put us back into the empty picture generated by this <u>limited world of science</u>, we've introduced a controversial thought. We argue all our experiences of the world are caused by systematically processing energetic events occurring at the molecular level. These events are by their very nature the same class as our experiences and emotions concerning the world. What these simple molecular scale experiences lack is concentration, organization and focus. Our powerful emotional impulses are derived after enormous numbers of these tiny experiences have been ordered and combined. The channeling of the energy inherent in these experiences is achieved using the informational component present in these inputs. This is where the embodied mind's computational power comes to the fore. Finally, the focused pent-up emotional energy is released as an action on the world. The sum of all the energy inputs and outputs are equal and no physical laws have been contravened.

Will-for-life

Freya Life always channels this energy into actions directed at achieving its aim of survival and reproduction

Orin: Well of course, this statement is also going to raise eyebrows in the scientific community. Science no longer has room for any 'aims'. Any aim sounds like the middle age scholastic notion of a movement towards the ultimate cause; that is God. Since the 'age of enlightenment', they have rejected this notion. This is why neo-Darwinism cannot allow such willful aims, nor are they relevant because the universe is observed only in terms of exterior relationships between inanimate objects. It is indeed purely physical and therefore by definition aimless. We, however, are focusing on the interior events, the subjective state of the system.

Freya: This gets back to our argument that although it's safe to ignore interior events raised in inanimate entities, the same isn't true for life. This is because a self-replicating form of chemistry can only complete its cycle if it can both survive and reproduce itself. To evolve, this chemistry must pass on useful chemistry to its offspring. This tiny chemical aim has now evolved into my powerful 'will-for-life'. It's a will entirely missing from the inanimate clumps of matter composing most of the universe but it's present in the tiny fraction of self-reproducing chemistry we call life.

Orin: We reject the concept this will is externally provided by a detached mind somehow entrapped in this chemical system. This will is embodied in the very system itself. It is what a living system does, so that in every situation it knows what it is to do. If my ancestors did not have a will-for-life, they wouldn't be self-replicators and I would not be here to talk about it now. Serving this will-for-life is the reason for any emotional state developed by a living embodied mind.

Freya: The power of these emotional states aimed at fulfilling this will is the reason why animals do in fact struggle for survival. The Neo-Darwinists maintained the physical purity of their inanimate machines uncontaminated by what is to them, an inexplicable and unnecessary self-preserving force. As a result, not only do they find it impossible to explain why life became more complex, they also have no basis for consciousness. Their argument for complexity is pure chance, but without a direction provided by this will-for-life, chance will destroy it as quickly as it builds. Their only role for consciousness is an unnecessary machine-generated fantasy. In contrast, our will-for-life naturally emerges from the very operation of the universe itself. This means we've eliminated all these issues from evolution.

Of course, Max's argument will be <u>any process just does what it does</u>. You can't call it an aim. An aim is something a person might have, to complete a task, say complete a marathon. All these processes do is to convert 'a' into 'b' according to their inbuilt programs.

Orin: The marathon runner is a great example. She must use the given starting point of the race then apply all of her abilities to reach the end before anyone else. Just as happens with all processes, she attempts to produce a determinate output from selected inputs. Whether she succeeds or not is open to all sorts of external factors from the weather conditions, to the strength of the competition, all of which are beyond her control.

Science rightly denies that the universal forces they work with, like gravity, electromagnetism or even natural selection have a purpose. But we aren't requiring universal laws to have aims. When we talk about life, we are concerned with an internal striving to succeed in its reproductive task. This inherited 'will-for-life' is the aim lying at the base of all animal actions.

Animated chemicals

Freya: We've made real progress with our task of understanding how emotional states constructed by embodied minds are still grounded in the real world. The problem is, I now find myself as some composite entity, transiently constructed from events derived from different parts of my body. Max has argued that since the body is made up of lots of separate bits, chemicals, cells, organs and so on, there can be no whole self, just a collection of embodied minds. Fragmented bodies mean fragmented minds, but I don't feel fragmented, I feel whole and continuous.

Orin: If we could show our bodies form a single continuous structure, then we can argue the same is true for our minds. To do this, however, you have to show every chemical component of the body is changed because it is part of a whole. This goes against the powerful evidence of there being no difference between animate and inanimate chemicals.

Freya: Thinking about this problem, I believe this identity is only true because they're considering chemical identity. That is, the chemicals are composed of the same elements arranged in the same way. This is indeed true whether they're inside the body or not. However, there's a more subtle form of identity, which is about the three-dimensional shape of chemicals. Chemicals, especially the long chains of chemicals you find in biochemistry can fold into an enormous number of different shapes. This is called their secondary structure, and without having the right secondary structure, things like proteins and DNA can't function.

Water turns out to be the key component needed to make these chemicals fold into the correct shape. It links different parts of the structures together using weak electrostatic forces called hydrogen bonds. What these lack in strength, they make up in numbers. Water composes 90% of a cell, but not in the form of bags of slopping water, instead it's all in a highly structured form. In this way, water acts as the glue ensuring everything forms a tightly organized and mutually interacting structure. Therefore, although in isolation no difference exists between the animate and inanimate forms of chemicals, when organized by a living system, they are going to be in very different forms. Even cells are attached to each other through similarly weak interactions so such forces hold our entire physical structure together. In effect, all the chemicals important to life have a <u>different form</u> when part of this total living entity.

The other aspect of living systems is their being in a continual state of action, destruction and regeneration. They're always involved in performing some form of chemistry and this activity is all about a flow of electronic energy. This is true whether we are talking about simple biochemistry or the transmission of electricity by nerves. The entire living system is chemically off-balance and it can only retain this structure by continual repair and regeneration. There are no living parts not involved in this process. In Max's analogy of a cell being a chemical factory, it's as ridiculous as saying the factory is in the same state whether or not it is working.

Orin: The problem is when philosophers split the mind from the body, and scientists dissect the body into its physical bits. They forget we really are not a collection of different components but a unified whole. When you break it into pieces, you destroy the properties of the whole organism in just the way the properties of water are lost when you break it down into its component elements of oxygen and hydrogen.

United mind and body

Orin: So carrying on along those lines, because the body is a unity, then so the mind is a unity. This means it can integrate all the events occurring within its structure into a single unified experience. The more complex the body, the more data it must process in order to bring it to this unity. The results for us humans are intense, rich and complex experiences.

Freya: So, we are a whole, not a conglomerate of bits, a total unity of mind and body, form and function. I know it's true because it's what I find in my own experience.

Orin: The unity of body and mind means changes in physical complexity are the same as changes in mental complexity. The concepts of mental and physical as two different types of entity are based on our abstractions from the actual. You can see how the physical and emotional are interlinked with the way a positive mental state helps to combat a disease and placebo medicines are so effective at curing diseases. Just as a mental experience is caused by a system's physical change of state, the reverse is also true; a mental state will cause the embodied mind to change its physical state.

Freya. That's right it's "no mind without a body, no body without a mind" because both are just different abstractions of what is, in fact, a single entity! It's very difficult to keep this concept and not drift back into the mind-body split mentality. Of course, Max was very pleased to point out how using brain scans you can show a decision is made before the person is aware of it. So when we think we've made a conscious decision, in fact it's already been made. We just think we've decided.

Orin: Yes, this is an example of how scientists still live in the dualist mind-body split world, while on the other hand saying the mind is embodied. What they probably mean by this is some form of partial embodiment, just limited to the brain and the mind it contains. As Max said, scientists, and especially neuroscientists, have an entirely brain-centered view of the world. In contrast, we see the embodied mind present in all the structures of the body, every cell, every organ is a mind, and the entire system is a total unity. When you accept this, any decision to act was *mine* wherever and whenever it was made. Now an outside observer turns me into an object and breaks my decision into a sequence of sub-decisions occurring in a timed sequence, but to me as a unity of experience, the decision is mine whichever pathway provided it. Their exterior clock has no relevance to my actual experience.

Freya: It's another case of Max chopping me up into bits, then denying my unity while retaining his own identity.

Orin: This body-mind integration also saves us from the equally dangerous idealist view we can change the world by thinking about it. To change the world you must act on it and the only way to do this is through an output of energy-information in the form of a physical action. We might change our perspective of the world by changing how we evaluate it, but this will also involve a change in the processing of events requiring the making and breaking of physical connections. The dangers of pursuing a life of pure thought or a drug-induced state of euphoria is other life forms are focused on the practicalities of survival and reproduction and they will reprocess you!

Freya: From our unified mind-body view, the best physical solution to a particular situation will always result in the best mental experience. Conversely, the best mental experience will coincide with the best physical solution. As the experiencing subject, we aim to satisfy the physical needs of our bodies by bringing all the many components of the system into a harmonious whole. When we achieve this bodily state we feel happy and satisfied, if not we try to do something to restore the balance.

Life is going to exploit fully any inherited capabilities allowing this subjective aim of integration and harmony to be achieved. This naturally explains the evolutionary ratchet resulting in the accumulation of ever more versatile and complex chemical processes. What we see in the fossil record is just the development of increasingly complex bodies. However, these bodies are just the outward sign of minds able to enjoy an expanded range and quality of experiences.

Orin: Living embodied minds continually aim at producing this state of satisfaction. This means our subject-inclusive science can accommodate the concepts of value: bad, better, best because all are easily attributed to the quality of the mind-body experiences. This is impossible with the current scientific explanation, which being concerned with exterior universal forces, is unable to find any value in the objects it studies.

Single cause

Orin: When we boil it all down, the argument we are pursuing is based on an overarching philosophical and scientific principle, the need for everything to be traced back to an original event. It is no different from physicists trying to write the 'theory of everything' to describe how all the various <u>forms of matter</u> emerged from the big bang. We, however, want this theory to encompass our subjective experience as well. We are requiring just as the physical needs a cause, so does the mental and that both have the same origin.

Freya: This returns me to the motive for my search for self; a dissatisfaction with the view I somehow floated into existence out of nowhere and as a result have no functional role to play in existence except as some abstract afterthought.

Orin: This 'floating in from nowhere' is the concept of <u>strong emergence</u>. It is the idea that entities can come into existence with properties bearing no relationship to their progenitors. In this view, matter, life, mind and consciousness are all entirely new phenomena totally disconnected from went before. If something emerges, then it emerges from something with this potential built into it. Life, mind, and consciousness, are the result of the organization of particular forms of energy into ever more elaborate structures. All the elements needed for conscious life and the qualities of experience it delivers, are present in the basic forms of energy used to construct the entire universe. Where this energy came from and why it has the potential to take on such complex patterns and forms I leave for physicists to work out!

Freya: This concept of 'weak emergence' allows me to understand how my embodied mind is revealing qualities already existing. This means my subjectively experienced world of beauty and color is not a figment of my imagination. It was always present even before animals with our level of consciousness evolved. It existed as a potential waiting to be unveiled by a life form with sufficient processing power. It's exactly the same way the potential for a proton existed right at the start of the universe but this only became actual when the intense heat and pressure dropped. With this step, the potential for chemistry opens up, allowing the realization of life and consciousness. Our evolution from this initial creative event is so much like the way the potential for a tree exists in its seed, but the seed for the universe contains a potential for everything, past present and future.

The potential for my rich world always existed, just waiting for an animal with a suitable level of consciousness to discover and make use of it. This world was one of such enormous value to their survival, that when animals entered into it they rapidly filled it with their progeny and we are one of the results.

We've constructed a convincing argument for a continuous chain of cause and effect leading from chemistry to conscious life. We did this by seeing how the transfer of energy by chemicals is the transfer of information. These events are the experiences that when processed, provide the quality and power of our emotional experience. We can depend on physicists to take us from the big-bang to chemistry. This means we can trace the fundamental scientific principle of a necessary and single cause, right up to consciousness.

Orin: It might seem we took an anti-science track along some idealistic mind-over-matter fantasy, but in fact, we are not idealists at all. By requiring a body-mind link right from the start, we demand and find far more reality than either the realist or idealist camps ever conceived!

Chapter 8: Properties of Self

In which Freya and Orin link the more mysterious properties of self to its role in providing the center of stability in an ever-changing world.

Self-sustaining cycles

Freya: The problem with this discussion is I now don't seem to be anywhere! We've replaced a phantom being with another entity entirely at the beck and call of experiences over which it has no control. I come into existence only when processing events, but then just fade away when they stop. There's no stability in this form of self. Surely, I'm something a bit more permanent than that!

Orin: On the face of it there should be nothing stable since even at the most basic level, physics describes everything as dynamic forms of energy in a state of constant change. Of course, there are stable things out there and the question is how do they become stable. One of the most important stable entities for us is the proton. It is the sub-atomic particle responsible for holding clouds of electrons in their place and so giving elements their chemical properties. It underpins all the chemistry we use to build and maintain our embodied minds. Happily, the proton turns out to be an enormously stable structure with a half-life estimated to be greater than 10^{34} years. When you remember the

84

universe is only 15 billion years old, which is 1.5×10^{10} years, then you see the entire age of the universe is an infinitesimally small fraction of the life span of the proton. When physicists look at the source of its stability, instead of finding an unchanging solid point of matter, they find yet another layer of energetic particles in constant motion and mutual interaction. Now, under the conditions we find in the universe today, these sub-atomic particles are not at all stable by themselves. However, when they are brought together in the form of a proton they all interact to form a self-contained fully enclosed cycle of energetic events. It is this <u>dynamic cycle of events</u> keeping the energy tightly wrapped up in the proton.

Freya: So that's the big-picture in a nutshell! So where is this heading?

Orin: Where it is heading is at the most fundamental level, process stability occurs when energetic forms are in a dynamic self-sustaining cycle. Such an entirely self-contained process allows the proton to resist the impact of external events. The same is true for chemicals made up of a number of atoms. These are held together by electrons that are also in a state of constant motion. A stable chemical only forms when their interactions continually reinforce and support each other. In this state, a negatively charged cloud of electrons envelopes and balances all the proton's positive charges. Now, I am thinking a form of dynamic self-sustaining process would stabilize of our embodied minds as we process the continual flux of information-energy. This self-sustaining process is you, it is the self, the essential entity providing the center of stability around which the world events flow.

Freya: It makes sense because in this way our embodied minds remain stable but dynamic, even with the impact of an ever-changing external environment. Natural selection in combination with a bit of self-replicating chemistry has produced us human beings, which are in effect enormously overgrown chemical molecules. The problem this self-replicating chemistry has to solve is how to keep stable long enough to ensure reproduction. The answer is to build on the stabilizing processes operating at the molecular scale. Like atoms and molecules, to obtain any level of dynamic stability, you also needed to retain an internal self-reinforcing cycle. These cycles are not immune from change of course. If another highly energetic proton hits a proton then it is destroyed; for us it's being eaten by a lion!

So the origins of my self lie in these self-sustaining cycle of events found in the very nature of the most elemental structures known to science. Without these self-regenerating processes there would be no stability or reality to anything.

Of course, in our case the self produced by this self-sustaining cycle will be much more complex and rich. It's richness and complexity may well increase as you age so more side streams, nooks and crannies, what we call memories, could be incorporated in each cycle. At base, however, the self would remain this simpler underlying self-sustaining cycle. This is the actual unchanging entity retaining our being even if there are no external inputs.

Orin: The downside of this very complex self is you are rather short lived, a mere 8×10^1 (80) years compared with the $8 * 10^{35}$ years for the simple self-reinforcing process responsible for a proton's stability.

The market place

Orin: The next question to answer is how this advanced living person interacts with the world. You must remain stable yet still process the trillions of inputs received from 'out-there'.

Freya: This underlying unprocessed flux of energy in various forms is the base of my derived reality and is beyond my control. I use these inputs to serve my own purpose, either to avoid death, hunt for food or find a suitable mate. Can we ever really understand the nature of this basic world of raw inputs?

Orin: If our minds were unrelated to this primordial world, it would indeed be beyond our understanding. This is the case when science maintains the mind is an illusion, and philosophers saw mind and matter as two separate substances. With us, however, the mind and body have both developed from this unprocessed universe of raw data. We are genetically related to it therefore, we have every reason to believe we can understand it.

For this to be true, when we process our inputs taken from this primordial world, we cannot introduce anything new into the data. We might discover new properties or processes to derive a world of value to us, or evolve new senses to increase our data input, but we do not invent any data along the way. We must use and transform data in exactly the same way mathematicians transform data from one form to another using data-processing rules. These allow ordering, combining and contrasting vast data sets into a form we can use. This is what we mean by having a revelation of the world. Our revealed world is equally as real as the primordial data inputs from which it is derived, it is just processed by us into shapes we can use.

The most important aspect of our revealed world is all the opportunities and dangers it presents exist in three spatial dimensions. The events also appear in an orderly sequence. It is in this structure we move and hunt for food and mates. The old scientific view dating back to Newton sees this derived world as the absolute one whereas in fact it is abstracted from our world of experience. In it, space is seen as a matrix in which objects are located regardless of their usefulness to us and time flows regardless of what happens. Space and time are a sequence of blank canvases existing even when everything is removed from the picture.

In the early 20th Century, Einstein asked if you removed all the objects from the universe, would this structure remain. With Newton it did but Einstein realized the structure of space and time is intimately involved with the presence of massive objects. Massive objects such as suns and planets all curved space to produce gravity and time. Then things got even worse when another group of physicists started to look at what happens on the very small atomic scale. Their quantum mechanics very accurately described the way atoms and molecules process packets of energy. The problem is time and any sense of reality disappears. In order to make this world real, it seems to require someone to experience these events. This is often interpreted as a self-conscious person. No one has yet found a way to reconcile the two different theories.

Freya: I guess we can be of some help there because in our view of reality, a self-sustaining system of any simplicity is a 'person' able to experience events. So we don't need something with the complexity of our minds in order for there to be an observer.

Orin: That is an interesting thought! The fact is we need to strip away the advanced revelation of the world resulting from our extensive data processing. We can then describe how we process this underlying world described by quantum mechanics to generate the everyday world of space and time Newton describes.

Freya: From my perspective 'out-there' is a place full of potential experiences, not of objects or other systems or forms of energy floating in a 3D world. The 'out-there' only becomes my world by processing the sorts of data most useful for my survival and reproduction. This would mean there would be no 3D world if we didn't have the tools to perceive it as such.

Orin: We cannot even have an 'out there' if there are no other 'persons' or atomic energy processors to input and output events which in turn cause this flux of events. Physicists speculate to underpin a universe of electrons and protons we need whole hosts of dimensions. They are, however, inaccessible to normal states of matter forming the basis of our embodied minds. This means we can ignore them and for practical purposes, we can take the residual three spatial dimensions as the first stable genetic platform from which the chemistry of life evolved. The evolution of the universe up to that point is in the hands of physicists and their immensely expensive particle colliders.

Freya: So we can agree, for me to have any experiences there has to be an 'out-there', a public space where I am able to take part in conversations with the other persons. They are all busily chatting away and exchanging, processing and responding to information-energy. From this noisy marketplace, I gather all the gossip and by applying a whole raft of information processing steps, I reveal a world able to serve my private purposes.

Orin: Perhaps we can get an even better understanding of this marketplace if we consider how a simple embodied chemical mind would use it to derive its own world. How would a bacterium perceive this market place and negotiate its way around it?

Freya: Bacteria must experience the world based purely on simple events caused by the pushes and pulls of chemicals dissolved in water. After all, that's all they can use to derive their primitive world. Even so, they know there's something out there and if possible will move to get it. I'm thinking of those bacteria with flagella, little tails they can spin allowing them to swim. If presented with a source of food they will swim towards it. They know the direction to go and get what they want. So, perceiving objects of value to survival and reproduction and getting them for your own is not something requiring a high level of mental processing and consciousness.

Orin: It does not mean they have space revealed to them in the richness we do, but they have some ability to measure the quantity, direction and distance of a food source and to move towards it.

Making it my world

Freya: To measure the quantity of food is easy; all you need to measure are the number of receptors being activated. The bigger the stimulation the more food there is. But, if the bacterium is to determinately swim towards food it must get information concerning its location. Now that's easy as well because as it moves towards a source of food, the receptors will fire more often and the excited state of its embodied chemical mind will increase. It simply has to go where the excitement is the strongest.

Orin: The problem is there must be some way to compare the present state with a previous one. This means even the simplest of life forms must have a memory. Somehow, it would need to contrast its present intensity of the experience with one in its recent past and take appropriate action. This is a real problem.

Freya: Well receptors on the surface closest to the food source would fire more often compared to others; so the chemical excitement would be sort of 'off center'. This imbalance to the system's base line operation would only disappear when the food surrounds the organism. Even then, its internal excited state would cause some physical change in shape compared to the one it has when there is no food. When it reaches the center of the source, this chemical excitement would entirely enveloped it.

Orin: Yes you're right! Thinking about it, all chemicals are affected by the close proximity of another so a resulting force is felt and movement occurs until it reaches equilibrium. Oils are rejected by water and move away to form a surface film. While under this influence, the shape of the chemical's electronic field will be slightly distorted. It must be under some internally felt strain. Whether it requires a large or small adjustment to the electronic structure, every part of the chemical will be subtlety changed in some way. When the external force is removed, it will spring back into shape, just like an elastic band.

Freya: There's the answer then, a primitive organism feels the presence of other chemicals in its environment based on their ability to cause internal <u>physical-chemical strains</u>. Living cells can amplify chemical signals a thousand fold using a cascade of chemical reactions. This means a tiny chemical signal from the environment can cause a massive change in the interior state of the cell putting it under a physical strain.

Of course, depending on the receptors activated, the quality of this physically felt strain will be different. This means the embodied chemical mind can evaluate the strain in terms of its effect on achieving its reproductive aim. It will have to resolve the forces and make process decisions on how to respond to them, to resist or to move towards them. These internal responses are going to be honed to near perfection by natural selection.

Orin: Importantly, the strain felt by the system will have a direction related to the location of its external source. This strain will be experienced as a vector force having both magnitude and direction. The practical effect is without any revelation of space or any concept of time, even the simplest of organisms can effectively locate objects in its environment, assign it a value based on the receptors activated, and take the appropriate action.

The greater the range of receptors and the more advanced all this processing becomes the more qualities are revealed. This in turn helps to classify data sources and differentiate one source from another. We can begin to see how data is handled by an advanced embodied mind using a method called 'continuous event processing'. This is the technique used by information-technologists to handle vast amounts of data arising from say the world's stock exchanges. There is no storage of raw data, so no long-term memory is required, it just flows through the system.

However, when a previously set event occurs, say a sudden change in a value of a stock, a whole host of higher order data processing systems swing into action.

Freya: That's interesting because a similar method is used to explain how, given the vast amounts of <u>visual data</u> being processed, animals can so rapidly respond to threats.

Orin: The next issue is how to present the processed results in a fast and efficient way. Whereas a bacterium may vaguely perceive its environment as areas containing chemicals of more or less value to its purpose, we experience a high definition world of objects over-written with qualities telling us about how valuable they are. We, and probably all highly evolved animals, achieve this by <u>projecting our valuations</u> back onto the sources of the stimulus. This is how we see a simplified but 'color enhanced' picture of the world. One in which we experience the apple as both red and sweet. It is such a good presentation we think these are the properties of the apple itself. Even so, these are not just abstract images but processed experiences. These are powerful influences pushing and pulling us into action just like the chemicals present in a bacterial world.

Freya: That explains why we aren't just presented with a picture of the world of objects, but one full of power to cause us to respond in some way.

Orin: The ability to project our valuations back onto the sources of the raw data can happen because the inputs we experience are in the form of vectors. This means they contain information about the location of the objects producing the physical strains we feel. This can be sound, light, taste or touch. They all come from somewhere and depending on the sort of input, we can more or less locate the origins of the feelings they produce. The linear propagation of light and sound make these particularly easy sources of information along which we can project our valuations of color or pitch. In contrast, we must come into direct contact for taste and touch to work. This means we project values such as hard and soft, sweet or bitter, right back onto the objects. We are so good at this they seem to be properties of the objects themselves. The process is one of objectification of the world.

It is through these primary qualities all life claims ownership of the environment; we make it ours by selecting only those aspects useful to us. We turn something not ours, that is the experiences caused by the inputted raw data, into forms of energy that are of great value to us in our struggle for survival.

Freya: The important point here is without our processing of this data, there would be no color, sound or smell. These properties of objects are all values we derive from the massive volume of inputs we process. This is why we are always at home in the market place!

Orin: To me what is interesting about the projection of properties is that we can put a different twist to the ancient philosophical concept that we see with our eyes rather than through them. The 'emission theory' was based on the concept that beams of light issue from our eyes and this illuminates the world. Science of course says otherwise, we use our eyes to collect light for processing. But if we interpret 'illumination' as making the world meaningful, then the eyes also allow us to project a set of privately derived valuations back onto the source. So, we do not see the world through our eyes as mere patterns of light energy, but as one of meaning and value.

In and out of time

Freya: We have described how living entities take events from the buzzing public space and make it theirs, but what about time itself? Is time also given? Do we take ownership of it and make it our own in some way? After all, objects we desire are somewhere else in space but we are all at the same moment in time, as near as makes no difference to us anyway.

The other fact is our subjective time moves at all sorts of different paces. Unlike the clock time of science, subjective time is not uniform but event related. I guess if there were no events for me to process, there would be no time.

Orin: You will find even the time described by modern physics is event related and of course, runs at different rates depending on where you are. Time slows down in the presence of gravitational fields. This means the atomic clocks on the satellites we use for navigation are corrected to compensate for the time on earth running slower than the time on satellites. These corrections are tiny, but if they weren't done we would find our satellite navigation systems showing us to be further and further away from where we actually are. Of course, the slowing down of time in the presence of gravity meets its ultimate at the 'event horizon' of a black hole. Here time is literally frozen because at this surface there are no events.

We have another aspect of time demonstrated at the other end of the scale. At the sub atomic level, all the events are entirely reversible and there is no need for time to enter into the equation. So at the nucleus of an atom, where the protons just oscillate between states, it is not the case of events being frozen, it is that being a self-sustaining cycle, time is irrelevant. They are in that sense out-of-time or timeless and not in-time.

Freya: But in the world we live in, the events are not perfectly regenerative. We never see the events in our life repeated but they flow and change.

Orin: So we have these external inputs continually changing, evolving and never repeated and we say these are in-time. But we also have examples of another form of process where events are continually repeated and so timeless. Finally we have examples where events are frozen and unchanging.

Freya: You're going to need to explain what that actually means to me.

Orin: Think of a world consisting of completely stable atomic energy processors networked together by an exchange of energy-data. In this world, the forms of the energy-data are always changing, but the energy processors remain the same. These stable processors experience a stream of events caused by certain forms of inputted energy. These inputs cause momentary perturbations to their self-sustaining cycle but are not strong enough to change the cycle permanently. In effect, the external data is just shrugged-off, leaving the processor unchanged.

Now think of us; we are also such a system. These events act as disturbances to our self-sustaining process and we say this flow of events is the passage of time. For us, time flows as a continual, never to be repeated stream because vast numbers of events in infinitely variable patterns are assaulting us. So, although our 'self' remains the same, the patterns of experiences we deal with never repeat so we never cross the same river twice.

Freya: The good thing about this interpretation is the passage of time doesn't require my presence. The universe is full of energy-information processors and these are always changing the nature of the inputs and events I process. I contribute to the flux as well by my responses. All the same, time is felt subjectively because I'm the one experiencing this flux against the stability of my own self that is somehow outside of the flux. But, surely that would mean I am also outside of time!

Orin: My embodied mind is a single integrated whole with no parts. This means there cannot be any space or for that matter time in my being. For this reason alone, you can conclude I do not exist in space and time. Now think about space. Because we are a unity, we remain at the focus of all these inputted chemical and physical strains and the source of the values we project onto other objects we find around us. We both agree we are integral to the functioning of our embodied minds, but where exactly are we located within it? Often it seems to be here behind my eyes in the brain, but it can be in my stomach, my painful back or out there with the hammer in my hand or even at my destination along the road ahead or in the story I am reading. Therefore, I am not just in my brain, but I am wherever the focal point of my action is. Wherever I am, this is my center, from this location all events converge and all my responses emanate.

Freya: It explains why 'I', being outside of space, and time never actually move. Say, I go away on holiday, my self remains where it's always been. Sure, the new environment brings new inputs to me and I'm required to take new actions, but I remain at the same location. All that's happened is the world has changed around me. The same with time: you grow up, your body is continually renewed, moods come and go, but you don't seem to change in yourself. You feel yourself to be the <u>one constant existence</u>, although everything around you has changed.

Creative progress

Freya: Max will argue we've just described a body clock, a mechanical timing device allowing coordination of the electro-mechanical device that is life. What then, do you think science is describing when it talks about the time measured by clocks?

Orin: Science replaces our subjective clocks moving at different speeds with an exterior clock about which we can agree. We all experience the overlay of repeating patterns such as the daily cycle of light and dark. We have out-there something like our interior timeless cycle, but now it is something everyone else experiences. This means we measure the flux of events against something shared. If clocks were caught up in the flux of change, they would be useless. The atomic clock is such a great measure of time because its tiny oscillations are unaffected by the exterior flux of events. In effect, a clock is a timeless fixed cycle of events against which the flux of events are measured. The paradox is Max's clock can only measure time because it is itself timeless. So even if the self is thought to be a mere clock, it would still have these extraordinary properties of being outside the passage of time. I am, however, much more than a clock. Unlike this machine, whose job is to serve me, I am full of my own purpose and the flux of events have real and personal meaning.

When the subjective experience is removed, you have a universe without purpose. This is the picture of a purposeless decay delivered by science. The decay occurs because the energy processors are inefficient and the energy in the universe will slowly degrade. Eventually there will no longer be enough of it for any processor to use. The simple atomic processors would still go on in their internal self-sustaining cycles for an immense amount of this physical time, but there are no external events and so no experiences. This is what science calls the 'heat-death' but we might call the 'event-death' of the universe.

Freya: On the other hand, we recognize the energy processors themselves do have a direction and a purpose, even if it's only to process the energy inputs according to their inbuilt programs.

Orin: In fact it is the failure to complete their processes that results in the evolution of ever more complex structures. The evolution of energy processors occurs because in the present state of the universe, we have such an abundance of energy forms resulting in highly complex and often unpredictable interactions. Energy processors can only resist change but cannot avoid it completely. If all things were totally wrapped up in themselves and stable, there would be no creation of new entities and no changes in the state-of-affairs. Imperfections in their self-regenerative cycles means the processors can self-destruct or re-combined into different forms. With nuclear scale processors, this is the basis of radioactive decay. The other effect is external and happens because the energy processors cannot control the influx of energy from their immediate environment. This is just what happens when protons are forced to interact with each other in the center of stars. Their self-sustaining tightly closed cycles of energy are forced open and there is an exchange of powerful nuclear forces. The new element helium with two protons sharing the same nucleus is a more stable adaption to this environment than hydrogen with its single proton. It was the continuation of such processes of nuclear evolution that produced the range of chemical elements needed for life to evolve.

The second factor is that the merging and recombining of these processors into ever more complex structures results in a more complex response to energy inputs. We can predict how hydrogen atoms respond in just about all circumstances using quite simple mathematics. As these chemical elements get larger, however, their processing of energetic inputs also get more complex. There are then many more ways for these elements to combine into molecules or respond to energetic inputs. One of these molecules was, of course, the self-replicator that opened up the pathway to our evolution. We are just at the highly complex end of this evolutionary process.

Freya: From this perspective, Darwin's theory is then a specific formulation of a more general principle of evolution. For a system to evolve new adaptations to its environment, it must be in an open exchange with the environment. It's because all these processes can potentially use each other as inputs that this form of creative movement is also destructive. We destroy our food so we can maintain our existence in the presence of the overwhelming power of all this energy-data we process. But, the only reason we have this problem is because we need to achieve our reproductive aim at the expense of other living forms.

Orin: That process of destruction permanently

changes the environment. While a cycle has no direction, this iterative process does, and the direction is always caused by an individual energy processor attempting to satisfy its own process aim. It fails, succeeds or evolves into a new system able to succeed. This new form will then dominate the environment altering it forever.

Our very existence means the universe will never be quite the same. We have left a track, an indelible one fixed in every subsequent created entity. In this way, our outputted actions become immortalized as well as setting limits for the future creative movement. The range of possibilities the universe had before our actions on it are now fixed in a certain way because of our actions. This is the past. We exist at the creative edge of this movement our actions have helped to create. This is the moving present, which can only be experienced as such because our being is unchanged by it. This to my mind is where we find the arrow of time and why there is no going back because all potential futures have become the actual one. The world moves on to a new ground because of what we and everything else does, whether we like it or not. Life's job is to regenerate itself from this new position and in so doing it changes the nature of the ground. The process of continuous creation is unstoppable.

Freya: The natural world we animals live in is in a constant state of change largely because of our own actions. This action produces results not always to our liking either. But, just like the proton, life also needs to carry forward its previous states otherwise it would have no stability. Animal survival requires self re-creation in a ceaselessly changing environment. The selection pressure is for creativity as well as adaptability. With humans and the more advanced forms of animals, we see an increased emphasis on these creative abilities allowing us to discover novel ways to solve the new environmental problems we have in fact created.

Orin: Where this unstoppable flux and the creative progress it drives might end is totally unknown. This contrasts with physical science's conclusion that the end is known; the total destruction of everything. This conclusion occurs because science focuses only on the transmission of energy between processors. Therefore, it can only show a valueless empty husk of space-time decaying into nothingness. As a result, it imparts a feeling of gloom and hopelessness causing such a negative reaction against it. In contrast, when life scientists start to emphasize the living self, we see the universe as a seed husk from which complex highly aware entities have evolved. We have no reason to conclude this process has reached the end. Much more complex embodied minds than ours may well evolve from us, or may already exist in other parts of the universe. We cannot discount the possibility these may exert a great deal of influence over the future evolution of the universe.

The unity of self

Orin: The technical challenge our embodied minds have solved is to bring the vast flow of simple experiences into a <u>single complex one</u>. It seems to do this by bunching data streams and enhancing contrasts then projecting the resultant output onto the original sources of data. These processes present us with a settled ordered world of objects with certain properties. It is, however, also one full of potential experiences and to make these actual we must focus attention on these objects. As we do so, the data handling systems zoom-in and bring more and more detail to our attention while still keeping the periphery under surveillance. While all this is going on, we remain the source of stability that can enjoy this flow of events yet remain unchanged by them.

Freya: The problem I have with basing the model of self on the simple cycle of events seen in elemental particles and molecules, is their cycles don't require an input of energy to keep going. Whereas it 'shrugs off' external events, our self-regenerative component must use the energy of the inputs in order to operate this self-sustaining cycle. What's more, our consciousness isn't present all the time, for example when we sleep or are anesthetized. To recover consciousness requires a highly energetic powering up of all sorts of complex processes.

Orin: Perhaps a better model is to imagine the fully conscious 'I' as some uniting field only brought into play by the continual supply of energy. It works in the same way a field of electrons unifies any chemical, only now it unifies the entire massive 'molecule' we call our person. The structure of the field would then become more enveloping as we move into more and more complex states of awareness. Stabilized wave structures are often generated by complex biological systems. Perturbations to the structure of this unifying wave caused by exterior events would be our enjoyed experiences.

Freya: When unconscious, our unifying embodied mind is not there and the various components carry on in their own manner. Without this unifying self and its drive to experience the world, the parts would rapidly lose their integration. Just as soon as a full regenerative cycle is established, then the fully united 'I' pops back into existence and we get on with fulfilling our goals. In that sense, our consciousness is acting like a clock coordinating all the systems and bringing them into line. To me, it seems as if I never lost consciousness and because I am outside space and time, I haven't actually gone anywhere either. If one of the key components fails, 'I' can't re-establish the full cycle needed for consciousness. Without my unifying presence, the body becomes a collection of component parts, is unable to function and so dies.

Orin: The interesting thought is when you dampen down all the exterior flux, you are left with the raw person who is the center of this pure self-sustaining field. Then, perhaps the timeless nature of our being comes to the fore. What T. S. Eliot describes as the 'still point of the turning world'. Interestingly, by using methods to suppress our continuous stream of thoughts, this is precisely the state <u>meditation</u> is intended to bring about. This brings consciousness into entirely new states of awareness. Perhaps the projection of this experience of a timeless state onto and entity called God is the origin of our religious beliefs. Of course, unlike God, who is eternal, this timeless state is transitory. The power of the flux is such that it will eventually overcome the self-sustaining cycle. We are in the end just temporary stable eddies in the flux, although when in this state, we are something quite extraordinary.

Freya: What we haven't done is to split this self from the body. This is because the wholeness of self and the potential for it to move towards these 'timeless and space-less regions' is only possible because the self is maintained using the flux of energy-information derived from the world. All the same, we do seem to be entering some very mystical areas here, and if we go any further, I'm sure it will confirm Max's worries that your philosophical inquires will lead me into the land of spirits and fairies.

Chapter 9: Reflections of Self

In which Orin and Freya discuss how consciousness becomes self-consciousness, words have power and we become free and civilized.

Self as an object

Orin: Our explorations into the origins of self have focused on the emotional consciousness we humans share with all animals. We now need to consider the state of being called self-consciousness. The consensus is that this is the unique ability responsible for the huge divide between us and other animals. This may be true, but we need to show it as a further development of consciousness. If it were an entirely new state of being, we would again have created an unbridgeable gap between reality and us.

Freya: We are not born self-conscious. Infants only pass the simple test for self-consciousness, the ability to see an image in a mirror as your reflection, after about 18 months. After my daughter reached this age she would stand in front of the mirror trying on all sorts of dresses and evaluating which one made her look the prettiest.

Orin: To pass this test, we must project our identity onto an image of ourselves. Although it might have profound implications, it is not such a big step because we project our subjective valuations such as color and taste or to objects all the time. Even so, what a strange thing it is for the conscious self. To see an image outside of me as representing myself must cause a profound change in how I see the world. Once I've seen myself in an image I become my own object and can project all sorts of properties onto myself. I can for the first time see myself as others would. I can shift the view of my actions to a position outside myself and think of the consequences of consequences and so on. The reflections can go on to infinity and I lose myself in them. These are the issues with being-in-the-world and the problems in regaining authenticity that have been so well explored by the <u>existentialist philosophers</u>.

Freya: From a practical point of view, you can now generate a 'theory of mind' and see other beings might also have minds that think like yours. The trouble is self-consciousness and having a '<u>theory of mind</u>', is not altogether unique to humans and tests have shown other animals share this ability. It's not even confined to our close relatives such as the great apes and social animals such as dolphins. For example, if a raven knows another is watching it, it will try to conceal the location of its food cache by pretending to store the food at another location. This is interpreted as the bird projecting its devious mind-state onto another bird and so knowing if given the opportunity, it too will come along later and steal the food.

Orin: The other way humans seem to be unique is we are able to use reason to determine our actions. This requires us to override our instinctive emotions with ones produced by this reasoning process. Our instinctive responses are fast and get us out of trouble without any need to apply reasoning. Reasoning, however, is more reflective and can be applied after an instinctive event. We use it to consider if there might be a better way to handle that situation. It is a way to free our future actions from instinctive ones. If these reasoned actions are superior in terms of survival and reproduction, then any animal able to reason will be selected. It turns out reasoning is indeed a very powerful tool. In order to reason, however, we need to think in symbols not emotions, and to do this, we need to have a symbolic language in which to think.

Freya: Language is not something unique to humans. We see its roots in the way animals communicate with each other using sound. For example, an animal's alarm call is a package of sound-energy structured in such a way that after processing it raises a feeling of danger in other members of a family group. Plants can also alert each other about say, an attack by a fungus, by emitting a chemical into the air that will switch on various defensive mechanisms. In fact, the whole natural world is in a state of communication using sounds, scents, and vision.

Orin: You are right; it does not matter whether the communication is via chemicals, sounds, or a written word on a sheet of paper. An input of energy; be it chemical, sound or light is always made, which is then processed as data allowing decisions to be made and actions to be outputted. Words, even though they are symbolic structures still act in the same way because provided you share the same decoding processes, they have the power to make you react in an emotional way. I remember an old rhyme we were taught as children, "sticks and stones can break my bones but words can never hurt me". I always wondered why it was not true. It is because words are not abstract symbols at all, but conveyors of enormously powerful emotions.

When self-consciousness allows us to see ourselves as objects with properties, we can observe our emotional states from a distance as well. We see they only form in certain circumstances. If I see my rival then I feel jealous and I want to attack him. Our ability to observe ourselves as if we were someone else allows us to recognize chains of cause and effect. We see how our own embodied minds respond to circumstances in predictable and logical ways. We discover a pattern of combination and suppression of emotional forms we call reasoning. Our language being emotionally derived also models these logical structures and so reasoning with words becomes possible. Finally we extract the logical forms from the language so the emotionally felt responses to a predator of fight or flight can be generalized into if x is true then y else z'. The logical operators such as 'not', 'or', 'and' are all present in the language of emotions. We have discovered logic and reason within our world. I say discover, not invent because rational responses to any situation are an essential part of survival and without it, we would not survive.

Freya: Language is also important for our social development. It's by sharing a language we cement our society together. It brings us into empathy with the experience of another person. It's as simple as pointing to an apple and saying 'good' or at a sour berry and saying 'bad'. We can pass on the likely experience the other will have when eating the object. Given our children are so slow to develop, we must have a very strong and mutually supportive social structure for them to survive. But again, there are plenty of examples where other animals form large social structures and this can only occur by means of communication. Insects are a prime example of this. So, there's nothing special about our dependency on social systems in order to raise our young or our ability to communicate.

People are always trying to find 'the factor' making us what we are. So it's say, self-consciousness, language, reason, the need to nurture our young over so many years, our large brains, or even the 'opposable thumb'. When we examine these individually, we find there is nothing unique about us at all. Instead, it seems the cause of the large mental and physical gulf lying between us and other animals is our possession of all these capabilities.

Thinking of it in evolutionary terms, we start with a mixture of these minor attributes each giving us a slight advantage. The feedback of natural selection then amplifies these minor differences. For example, mutations increasing brain size allow increased ability to put oneself in other people's shoes allowing improved communication and social skills. When this is combined with an opposable thumb, we can pass on weapon making skills to our kin. The better teamwork, communication skills and weapons allow us to bring down bigger game and to defend it from other predators etc. etc. Over a few hundreds of thousands of years of natural selection, these minor enhancements lead to modern persons with their advanced language, reasoning, food production and social skills.

Free thinking

Orin: In our descriptions of how life and consciousness evolved, we describe our actions as a computed output based on a set of inputs. As you pointed out, this means we have destroyed the concept to free will.

Freya: Well there's freedom because although what's in the environment is fixed, how I use it to serve my will is undetermined, or at least only partly determined by these objects. In any case, the shear complexity of the process means the outcome is undetermined. It's well known that a system experiencing such a complex range of inputs has got to go into chaotic states. When this happens, we chose a path at random based on tiny fluctuations somewhere in the environment. So we're not at all determined by the past state of affairs.

Orin: But you are just admitting being pushed around by forces over which you have no control. I am afraid it is not at all the same as having free will.

Freya: I guess you're right; any action an animal

takes is with the aim of self preservation. That's even true if it sacrifices itself, like bees will sting you to protect the hive, even though it kills them. It will always be for the good of itself by protecting its genetically related offspring.

Orin: My theory is that <u>freedom</u> is only possible because we can use our complex language to tell stories. Our primary use of language is not to make true statements, but to connect and empathize with others. We use language as poets; after all 'to be or not to be' is logically nonsensical. As poetry, however, it is very effective at making us understand Hamlet's state-of-mind as he contemplates suicide. It is because language is able to make illogical statements our minds can break free from our instinctive logical responses and socially provided solutions to explore other potential realities.

Freya: I see what you mean; using language, we can take the wings from a bird, fix them onto a man using wax, have him fly to close to the sun where the wax melts when he crashes to the earth. By playing games with words, we can conceive new things, and enter into the world of imagination where anything is possible.

Orin: Just so often, we can turn a potential reality into an actuality, such as a real flying man. Admittedly, to bring it about requires the insertion of a massive quantity of technology, but it has happened because the myth was so powerful and attractive to us we worked and worked on it until we got what we wanted.

The sort of freedom we are discussing is not

absolute freedom. It is bounded by the nature of the universe. There is no true thinking outside the box, because there is no way to escape from the box. What we can do is to uncover currently unpopulated potentials for existence and attempt to make them actual. We use reason to check a route exists between our current state of affairs and the new one. If it does and we follow this route, the potential becomes actual. The problem is, as with your Icarus example, the consequences may be disastrous. Whatever the consequences, these new actualities are all part of the continuous process of creation. All beings must adjust to these new state-of-affairs or become extinct.

Restraining instinct

Orin: The fact we can use reason to re-direct our instinctive responses is the essential foundation for our advanced civilization. We assume a person is able to repress their instinctive emotions and replace them with reasoned ones. In this way, they are expected not just to know how to behave for the good of society, but to actually do it.

The laws of the land, generated over thousands of years of practical experience, are instructions telling us how we must behave. Starting from the obvious socially disrupted ones of not stealing or murdering, we have moved on to being told when and where we can cross the street. The problem is you cannot write enough laws to cover every eventuality. This is where the attempts to instill moral principles become important. They tell you how you ought to behave towards your fellow citizens.

Freya: So it's these civilizing rules that 'in these circumstances you ought to do this or you must do that' become the potentials for our behavior But these are only possible because we have this ability to suppress our instinctive actions with reasoned ones.

Orin: As patterns of energy, these laws and duties have power in themselves to make us act, but because we have the ability to produce counter thoughts, they can be suppressed. The 'good citizen' doesn't do this but turns them into an actuality. My society wants them to become such a part of me that I will act by instinct rather than reason. A lot of education is to try and build these moral concepts so they become automatic and not countered by acts of free will. A person deciding to go against these imposed limitations on her freedom is a rebel. If it is a serious rebellion and you break a law, the state will restrict your physical freedom. You can end up in jail or even dead.

Freya: If I'm given a 'you ought to do this' conflicting with my will-for-life, I'm going to feel this as an emotional conflict. I will need to balance the pros and cons of suppressing a social 'ought', which will be for the good of the community, with a counter thought based on my own, and my kin's survival. I think my will-for-life is going to win out every time!

Orin: That is because you've accepted your animal nature. However, there have been periods in our own history where this animal side was denied because we were seen as God's unique creation. To show our special nature it was necessary that your actions were not guided by animalistic self-interest. By performing some act of self-sacrifice it was believed this superiority could be proved. The problem is you could never convincingly show such acts are truly free, or whether they are just an expression of an underlying animal self-interest. Even so, it resulted in concepts of honor, duty and citizenship that seem to be inconceivable in our cynical days.

Freya: In my view, in the best social systems the rules we learn won't conflict with the basic instincts derived from our will-for-life. After all, the only reason for having a government is to support our ability to survive and reproduce. The law should mutually enhance our family life and surely that's what the social contract between us and the state is all about. We would have to act against both our genetic and acquired social rules to give the ultimate example of free choice. The problem is this would just be acting in a contrary way and is very likely to have fatal results.

Orin: All the same, the freedom provided by our advanced self-consciousness means that you could use reason to counter instinctive responses. When scientists argue humans are purely machines unable to act contrary to their inbuilt programs, they are taking away the very foundations of our civilization. This means you can avoid all responsibility for your actions. It is certainly true our minds are dominated by a will-to-live. However, our reasoning abilities and highly developed language allow us to balance alternatives in such a way we can exercise free will; even if it is rarely attained. This means humans are able to perform truly good, as well as evil acts.

Poetry of science

Orin: We affirmed we are at least free to string words together in any order we like. The more pleasure they give us, the more we celebrate the storyteller. Now if this is so, what is to say this whole discussion is not just another piece of poetry, just a jolly good story to give us a bit of pleasure?

Freya: We are doing it because we believe we can discover the actual state of the world. Everything we've discussed so far is based on the belief it is possible to develop a better explanation about how we evolved into self-conscious animals without any mysterious steps.

Orin: The argument made by post-modernists is science is just another story with equal value amongst the rest of our myths. The creation myth of the Bible has an equal standing to Darwin's myth of evolution by natural selection. You have to be fair and realize there is a sound reason for this and as presently formulated, science finds this impossible to answer. The reasoning goes that objective science is about discovering the properties of objects. As we have agreed, an object's property is a projection of our subjective valuation. The belief that when science strips all this away we can still find something independent of the observer, does not stand up to careful scrutiny. However much you twist and turn the reasoning, the mind of the scientist is always the final component in any measurements it takes or observations it makes. If the mind is an illusion and we are all robots, which is the only conclusion materialist can make, then so are all the measurements it makes. This material world then vanishes into an illusion generated by an illusionary mind.

Freya: If this is so, then how can science and technology deliver so many useful and practical things that actually work?

Orin: This pragmatic response is the only answer you can give. A purely materialist position is based on the faith that the concepts we have of things correspond to the actual things themselves. What they cannot argue is that concepts are derived from actual things. This magical 'correspondence' is hardly a sound basis for you to construct your massive edifice of science is it?

The great news is we have no such problem. Provided you accept our basic concept that experiences are patterned energy which, after extensive processing produce specific emotions, then we are always in touch with reality. With this base and our 'bottom-up' approach to self-consciousness, everything we express in words is derived from emotions deeply grounded in simple molecular events.

The problem with our highly developed form of self-consciousness and ability to express these feelings as words is they allow us to believe we are disconnected from these chains of events. We think from a perspective outside of ourselves with words that seem to be disconnected from the reality of feelings. Words are powerful because as patterns of energy derived from actual experiences, our embodied minds can revert these words into an actual feeling. If we want to, we can allow these words to penetrate into our consciousness so we actually experience the emotions.

Freya: Look at how the high levels of abstraction used in 'financial instruments' were still linked to the reality of people needing to be able to repay the loans they were offered. When the shaky foundations were exposed the whole edifice came tumbling down and with it a large part of our pension investments. Thinking that living in a world of high abstractions means you are divorced from reality is amazingly dangerous. The fact we can be fooled time and again by this is because we are taught to see words and numbers as inert symbols rather than packets of elemental energy ready to raise powerful emotions and responses in us. We are processors of experiences, and words are immensely powerful emotion-laden packets of energy.

You can see why those scientists working with artificial intelligence have such a hopeless task. Because they still see consciousness as purely playing with abstract concepts, they believe they will eventually be able to make a computer aware simply by teaching it enough words and meanings and rules about how to combine them all. Taken in isolation from the embodied self-conscious mind that generated them, there is no way this simplistic approach can bridge the gap between reality and the words used to describe that reality.

Orin: Science is also differentiated from a purely imaginative story because the narrative is constrained to those things we actually find in the world, rather than to what we might wish to find. The scientific method ensures our concepts are derived from the directly experienced world. These are those stubborn facts we face as we recruit these given objects for our own purposes. Now, even though we describe them in the most general abstract terms as if their properties were independent of us, they are not. Instead, these properties are derived from pulses of energy transmitted from the object and felt by us as a pulse of emotion. The properties, however much we derive and enhance them by our subjective processing are still based on a real exchange of feeling between us and what we are observing. We are never the detached observers we pretend to be.

Freya: I guess the next step is an imaginative one. This is where we construct the most believable 'story' about why these properties are just so.

Orin: These are our hypotheses and theories, but again just as in a murder mystery, the form of the narrative must comply with certain rules. Science, has got to tell the simplest story that is self consistent, logically coherent and inclusive of as many observed facts as possible. Now that points to the next constraint; the test of their efficacy is in whether they can be used to create useful concrete actualities or predictions about some future state-of-affairs.

Freya: This gets back to the pragmatic test of our theories; they can be used to create something of practical value. I can use Darwin's theory of evolution to explain how a microorganism can grow resistant to antibiotics. The resistance is factual, but the concept that a small number of survivors were resistant mutants and these grew in their place, is the simplest story best fitting the facts. This is a variation on his story about why different forms of finches evolved on the Galapagos Islands. That's what I call a theory with breadth.

This isn't true of fiction or poetry or religious myths. Their power is in making us have some form of emotional experience based purely on the order of words or the images they produce. No concrete actual entity needs to be made, but this doesn't mean the emotions they raise in us are any less real than those raised by a piece of shiny new equipment produced by science and technology. If you want to test this out just pick up a book or see a movie and experience the emotions these images can cause you to have.

Doing science itself raises a whole raft of emotions in us scientists even though we try to pretend it's all occurring in some impersonal way. For example, after 6 months of work when I managed to successfully isolate my first gene, I ran around the lab whooping with glee then went out with my friends for some serious celebrations. When I later wrote about it in my Ph.D. thesis it became: 'Sequence analysis demonstrated the presence of the desired gene construct'. Without the enjoyment of these emotional highs there would be no science.

Reality rules

Orin: Just about everything our society has achieved in mathematics, science, literature, ethics and art, was obtained by combining these symbols and testing the powers they released. This is how we explore the potentials for creating new emotional states in ourselves and in other people. We do this by telling a good story or making a new anticancer drug. The novelists and the scientists are both having the same effects; they just call on a different set of technologies and skills.

Analysis of this highly abstracted world of symbolic thought and reason does not reveal an ultimate reality. It does, however reveal the underlying processes required to turn raw experiences into the reality of our world. Without these structuring rules, no events would occur because there would be no systems, simple or complex, unstable or stable, to experience and process them.

Freya: Then we could argue these processing rules form the basis of all reality.

Orin: Finding these rules is precisely what science, and mathematics tries to do by abstracting the patterns structuring 'pure' energy. These multitudes of forms of energy provide the specific type of experiences we have. Without energy to invigorate these forms, there would be nothing to experience. Just as there is no such thing as formless-energy, there is no such thing as energy-less forms. It is the combination of form and energy that makes the experience; it provides something out of nothing, order out of chaos.

Physicists think that if you add up all the positive and negative forms of energy, the <u>total is zero</u>. If we describe the primordial state of the universe as a ferment of positive and negative energy with no structure, it would simply collapse back into nothingness. The mystery is not just where the energy came from, but how this structural information was encoded. Another way to look at our evolution is to see it as the progressive unfolding of structured forms of energy into ever more complex forms. This mathematically and logically controlled process continues as the universe develops through various stages. We now see it divided and ordered into a multitude of separate entities each expressing different levels of awareness. So everything we experience from the color red to feeling happy and even our most abstract thoughts concerning the nature of reality all require the generation of highly complex patterned forms of energy from simpler forms.

Freya: As far as we know, our self-conscious states are the most complex forms of energy in the entire universe. The depth and breadth of the world we make real is all the result of our ability to control the formation of these complex forms of energy from billions of simply constructed inputs. These patterns contain an ever expanding horizon of entities, some actual, some from distant past and some with the potential of becoming our future.

Orin: If we were to imagine this evolutionary process continuing then there is no reason to think we are by any means the ultimate expression of this creative process. The ultimate form of this movement would of course be God. Whether this lies in our creative future, we cannot tell but if it were so, there would be an impact on the entire history of the universe. The thing about such an entity is it would know all the potential forms our futures can take. This is very much along the lines of Plato's concept of the mind of God containing the eternal and unchanging 'forms of things'. Their actualization depends on whether these forms are 'energized'. It is by our present actions we determine which ones will be actualized in the future. This means although all futures are known to God, they remain as mere potentials. God is concerned with our actions, because what we do now determines which set of potential forms will become real. God's concern with the world is then the source of our religious experiences.

Freya: What can't be denied is people do have religious experiences and I guess that's the only reason we have churches and organized religion today. The current conflict between science and religion is down to a few churches taking their creation myths as literal rather than poetic truths. The next problem is they keep telling us how to live our lives according to some tribal rules that no longer apply.

Orin: When you take away religion's claim for being the ultimate source of our knowledge about how the world works, and how we should live our life, then you are left with a core function; to bring people into this mystical experience of God. All the rest is just playing politics, and politics is best left to elected politicians, not to some priest or ayatollah who knows all the answers because God revealed them.

Bringing it all together

Freya: Frankly, I'm feeling rather shocked about how my search for self got us to this point. We've covered such a wide field of thoughts its time we brought all the strands together and see where we actually are.

First, I feel a real sense of achievement in finding how the self is an essential part of evolution by providing the center that actually cares and struggles to survive. This entity is the complex self-sustaining cycle of events providing the necessary center of stability. It is from this 'still point' I experience the world as a flux of emotions. The exact nature of feelings I experience comes from the selection and processing of the energy-information incoming from my local environment. This processing makes the world a place I'm completely at home in. In physical terms, this is the flow of energy that is structured to produce actions increasing my chances of survival and reproduction. These channeled forms of energy are my emotions. They disturb my self-regeneration and demand actions that increase the probability of my own regeneration.

Orin: Without this self, there is no point of reference against which these changes can be experienced. Without the flow of events caused by the input of exterior data, there would be nothing for the self to experience.

Freya: Psychologically, I now feel connected to everything else, yet I am still my own person. It all makes sense, my struggle is personal and my will-to-live is personal. In addition I'm free to think what I want, although I'm restricted in what I can make happen by what has happened in the past. That includes my whole chain of ancestors stretching back to the start of it all and not just life either, but to the very moment of creation itself. That makes us all very special.

Orin: What interests me is the way in which we can throw such a light on the nature of consciousness just by this small change in our description of things. By seeing that what physics describes as exchanges of energy-information are actually felt as tiny emotional events, all the mysteries of our consciousness have gone. The processing and uniting of these feelings into a unified experience of the world are readily described by the rules governing information processing. It just needs the selection, combination and amplification of these tiny emotions by our cells, organs and brains into one powerful emotion making us act.

Freya: To me the most important aspect is that our valuations are not just practical but pleasing. This encourages me into a world of potential delights, rather than retreating from a world of terror. Both beauty and struggle have a role to play in evolution. Evolution is just as much concerned with value as it is with survival.

Orin: You realize all we have done is to allow science to link up with what has been obvious to all people; we are in-the-world and of-the-world.

Freya: This gets to an important point about why we scientists have found it increasingly hard to connect with people. It's because our approach to life has been to treat it like a machine. The 'real science' of physics taught us this was the only way it could work. Since then we've generated all sorts of problems for ourselves, all because we can't find the observer needed to bring our universe into reality. Now we're proposing a simple solution; the observer is built into every operation occurring in the universe. Through the continuous exchange of information between energy processors, the whole cosmos can be seen as pervaded by a primordial awareness. It's just not structured and this structuring is just what life does.

Orin: Although this simple thought enlightens so much of the world currently closed to science, our colleagues are likely to reject it out of hand. It seems at first glance irrational and an attack on all the achievements of science.

Freya: All we can argue is its acceptance will bring science in from the cold and allow many of the important messages about the future impact of our activities on the world to be taken seriously. Scientists intent on reducing humanity to mechanism have caused science to become even more <u>alienated</u> from the rest of culture. Where people really live is in a world where our feelings and emotions have value above everything. There can be no relationship between the two polarized worlds we have generated. That's why whatever the consequences to our scientific credibility, we should work to bring the argument and our dialogue to as wide a range of people as we can.

The Origins of Self

130

Part III: Beyond Science

Chapter 10: The Discussion

In which Freya and Max discuss Orin's write-up of their conversations and Orin shows how their explanation of our consciousness parallels aspects of Process Philosophy.

Feedback

Orin: This short book is my attempt to bring all our debates and discussions on this topic into some form of order. Do you think I have captured all your thoughts and arguments and more importantly presented them in an interesting way?

Max: Well I'm not sure you've portrayed me in the best of lights. I've come out as the devil's advocate in all this. However, I can see you need someone to present our current scientific understanding about the nature of consciousness. Your use of conversations also makes it easy to read, and I think you've described the science part of it in a readily understandable and technically acceptable way. It's very good of you to spend your time putting this together for publication, but do people, scientists and the public really care about this sort of thing?

Freya: Well I think it's important to let people know they're not machines, and to explain to scientists why their science as presently constructed is limited in what it can say about consciousness. As it happens, the problem turns out to be a philosophical one. Even though we all are 'Doctors of Philosophy', the philosophical part of our education is virtually non-existent. So it's good to see philosophy brought into the open in a way even hard nosed engineers can understand.

Max: Well, that's the part where I start to have issues with the work. It gets even worse when you start moving into this 'metaphysical' stuff. Metaphysics just means pure speculative thinking to me. I don't think it has anything to say about science, in fact it doesn't seem to have much to say about anything useful at all. It's just one of those dusty old subjects mouldering in the backwaters of universities.

Orin: Well the term 'metaphysics' was first applied to the works of Aristotle. It just means the works he wrote after his 'physics'. In that way metaphysics is informed by physics, and there is no reason why physics itself cannot be informed by metaphysics. It is all part of the necessary interdisciplinary learning cycle. During the 20th Century, increased specialization caused science and the humanities to separate and this allowed basic philosophical mistakes to be made. I hope that in the 21st Century with the internet giving everyone access to information form all sorts of sources, we might be able to tie all the bits back together. That is all metaphysics tries to do, re-unite all the bits of science, morality, religion and aesthetics into one coherent whole.

Max: Well you and Freya certainly did go way

beyond science. By the time you got to the end you were speculating about life, the universe and everything, and you claimed our being 'out of space and time' was the cause of people's mystical experiences. You were both moving into the realms of fantasy and even your philosopher friends are going to think you've lost your bearings.

Orin: Well, the good thing is we have not invented anything new. Someone highly respected in both the mathematical and philosophical realms has done the work for us.

Process and reality

Orin: I'm referring to the works of Alfred North Whitehead. I think in many ways we have just re-discovered his concepts as described in his metaphysics entitled 'Process and Reality'.

Max: If his work is so relevant to the consciousness problem, why haven't we all heard about him?

Orin: He is well known amongst philosophers, but in order to understand his work, you do need an extensive background in the whole subject of philosophy itself. Even then, it is a notoriously difficult metaphysics to follow. The first edition produced in 1928 was full of errors and the corrected addition did not come out until 1978. This lists 22 pages of corrections[1]. Whitehead was also not one to bother selling his ideas to the public. For example, in the late 1920s he was invited to present his work in a

[1] Whitehead, A. N. Process and Reality Corrected Edition (Eds. Griffin & Sherburne) 1979 ISBN 0-020934570-7

famous series of lectures (The Gifford Lectures at Harvard University). The audience of 600 who eagerly attended the first lecture were so baffled or bored that by the second one, the audience had dropped to six. It seems only two lasted the whole ten lectures. Probably this is because the first part of the lecture was a series of definitions of his new technical terms. This is not a great way to grab people's attention.

Whitehead tells scientists their purely physical interpretation of the world is just plain wrong. They feel into the trap of adopting the 17th Century philosophy of Descartes who divided mind and body into two entirely different substances. Once this was done, there is no way that they can be reunited. As he says in his metaphysics:

'The disastrous separation of body and mind, characteristic of philosophical systems which are in any important aspect derived from Cartesianism is avoided.'[2]

Our inability to understand consciousness and to find the origins of self happened because philosophers interested in the nature of being have focused on the purely mental aspect of consciousness. Meanwhile, scientists have pursued the purely physical side of the body without reference to the subjective aspect of reality. As a result of this specialization, we now have a total communications

---

[2] P&R p246

failure. This is the problem Whitehead's metaphysics can help us solve.

For our point of view, his most important concept is what he calls the 'Ontological Principle'. His simplest explanation of this principle states:

> 'there is nothing that floats into the world from nowhere'[3]

Freya: Well I guess that's exactly the main principle driving my own search for the origins of self. We need a basis of self in evolution, otherwise we are left with a disconnected or even worse, an illusionary self that somehow floats in from nowhere.

Orin: His overall aim was to describe reality not in terms of objects with properties but of process. The world of objects with properties we think of as reality is the result of massive amounts of data processing by our embodied mind. This presents a picture of the world in the form he calls 'Presentational Immediacy'. This presentation of the data is so good that we think it exists independently of our processing. We have objectified the world of dynamic events to such an extent that it seems full of objects with fixed properties. We then think the statement 'sugar is white and sweet' gives us profound insights into the nature of the world. In fact, all we are doing is analyzing the subject-predicate structure of language! We independently arrived at the same conclusion when we saw the origins of our experience in

---

[3] P&R p244

physical events. He, like us, saw our conscious state to be the result of processing these experiences into a self consistent and unified presentation of the world. If you want to find the ultimate reality, it will be experiential events, not the derived objects that populate our world. Such events are the final realities because any further search for something more real will only reveal ever more primordial events. These he calls 'Actual Occasions' or 'Actual Entities'.

Creativity is the process by which novel actual occasions emerge. Novelty happens because an actual entity only emerges by unifying the inputs derived from other actual entities. The output from this novel entity can then enter into the formation of another one. The complexity of this process means the universe is not just a cycle of re-creation, but has the potential to make truly creative advances. Therefore, evolution is not an ephemeral side effect of physics, but becomes the ultimate matter of fact.

We realized the inputs are always some form of information exchange between us and another entity. He calls these external interconnections 'Public Matters of Facts' or 'Nexus'. His 'Prehensions' are what we described as vector forms of energy with size and direction and these penetrate us like arrows. The way the subject experiences these energy vectors are the private matters of fact he calls 'Subjective Forms'. The key to notice here is the subject and the experienced events are entirely bound together, there is no mind-body separation. There are however processes of integration and comparison deriving new subjective forms. These 'Modes of Synthesis of Entities in one Prehension' are also called 'Contrasts' or 'Patterned Entities'.

The final aspect is of course everything must have a definite structural form. These 'Forms of Definiteness' are perhaps the most mysterious entities because they exist whether or not their forms are occupied by actual entities. It is the concept we discussed that, for example, mathematical structures exist even though we have not yet found them. We discover these forms rather than invent them and the potential states that actual entities can assume he names 'Eternal Objects'. There then follows a whole list of explanations and principles defining all the ways the primordial world of chaotic and primitive events are processed to produce our ordered world.

Max: OK, you've made your point, I will accept that as a philosopher you have avoided the horror of actually inventing something new. Whitehead has done all the slog and created this 'metaphysical background' making you feel comfortable that your philosophical speculations about consciousness are on firm academic footings.

Panexperientialism

Freya: Now the union of mind and body required everything in the universe to be in some way aware. Does he address this very difficult concept? This, after all, is the key idea of how consciousness is linked to reality. It's also the most difficult one for anyone to accept.

Max: This sounds very much like the concept of panpsychism, the ridiculous idea that all matter is conscious, has a mind or a soul. Is this something you two really want to support?

Orin: Rather than 'panpsychism', the even more clumsy term of 'panexperientialism' has been applied to his concept. This emphasizes the fact that although experiences are had by all matter, this does not make them conscious. Consciousness is reserved for those living organisms that are able to integrate and channel these experiences into actions directed towards reproduction and survival. In any case, Whitehead never uses these terms; instead, he puts it this way:

> *'If we substitute the term 'energy' for the concept of a quantitative emotional intensity and the term 'form of energy' for the concept 'specific form of feeling' and remember in physics 'vector' means definite transmission from elsewhere, we see that this metaphysical description of the simplest elements (sensa) agrees absolutely with the general principles according to which the notions of modern physics are framed.'*[4]

He, like us, believes we can base our consciousness in energetic events without overturning the enormous advances in physics. We can have panexperientialism

---

[4] P&R p116

while still leaving the physical world intact.

Max: But the 'modern physics' he refers to has made even further advances since his time.

Orin: Agreed, the physics he is referring to was that of the late 1920s, but this is when the two great theories of quantum mechanics and relativity emerged. These still form the corner stone of physical science. Since then, much of the work of theoretical physics attempts to reconcile them into a single coherent theory. Something they have failed to do. It is outside my field, but I think it would help them shed some of their wilder speculations if they were to put the conscious subject back into the universe.

Freya: What other aspects of his metaphysics seem relevant to our discussion?

Orin: Whitehead often uses the term 'philosophy of organism' instead of 'process philosophy'. His emphasis is on seeing life as being in-process; an agent that is part of a society of other organisms cooperating and competing in a yet wider ecological environment. This is in total line with our present day systems view of life and evolution. The failure of biology is its inability to explain how a purely physical machine can give rise to consciousness. Just as he argued when you accept the concept of panexperientialism, the problem just goes away.

Max: So, you find it convenient to accept his philosophy to explain certain facts-of-life, but that doesn't make it right.

Orin: I suppose in the end the justification for

accepting any concept is that with it, we can achieve a more fully consistent understanding of the world and our place in it. As Whitehead says, the aim of his metaphysics is to produce a:

'coherent, logical, necessary system of general ideas in terms of which every element of our experience can be interpreted.'[5]

Freya: For me this boils down to the pragmatic fact that when an animal really struggles for its own survival, the evolution of fully conscious beings becomes inevitable, not an extremely unlikely event. We must have an aware mind and physical body in co-evolution because we can't have one without the other. They just aren't separate entities.

Orin: A quote from Whitehead confirming that you and him are on the same track:

'I now state the thesis that the explanation of this active attack on the environment is a three-fold urge: (i) to live, (ii) to live well, (iii) to live better. In fact, the art of life is first to be alive, secondly to be alive in a satisfactory way, and thirdly to acquire an increase in satisfaction.'[6]

God and the world

Orin: Now we need to face the 'elephant in the room', the God Whitehead continually refers too.

---

[5] P&R p3
[6] The Function Cf Reason by Alfred North Whitehead, p5: Princeton University Press 1929

Max: Knowing your interest in religion, I thought you were going to bring God into it somewhere. The only redeeming feature of your whole exposition is you link consciousness right back to a single event, presumably the Big Bang. As you said, this has removed any mysterious steps from the whole evolution of life. Now, I thought, Orin can finally join us atheists.

Orin: As a philosopher, Whitehead uses God to complete his metaphysics. So just like Plato, he uses the 'mind of God' to both generate and locate his eternal objects'. Without these structural forms, there would be no order in the world, just chaos. These are not just the mathematical structures of particle physics and quantum mechanics either, but include all the experiences and ideas we have of the world. All these need to be embedded as potential ideas at the moment of creation then revealed by our embodied minds as required. If not, they must mysteriously emerge from nowhere.

Max: Now that just makes God some abstract entity sitting outside this creation. He just sets up the initial state of the universe and all the potential states it might take, then kick starts it all at the Big Bang. Then sits back and doesn't care a damn about what goes on in it. He is immune from all the suffering and killing involved in the process of evolution and the struggle for survival.

Orin: God is always involved because creation is continuous and ongoing. God is also a person selecting what to incorporate into his own being:

'The consequent nature of God is his judgement on the world. He saves the world as it passes into the immediacy of his own life. It is the judgement of a tenderness which losses nothing that can be saved. It is also the judgement of wisdom which uses what in the temporal world is mere wreckage.'

God's active and continuous part in creation is actually a necessary part of Whitehead's metaphysics:

'In this way, by the recognition of God's characterization of the creative act, a more complete rational explanation is attained'[7]

Max: This vision leaves me cold. Not only is it likely to alienate those professing a fundamentalist faith derived from the Judeo-Christian-Muslim tradition, but also provides us atheists with yet another example of people's primitive need to believe myths rather than facts.

Final thoughts

Orin: If I were you Max, I would just ignore this God aspect, and view it as the philosopher's method for grounding metaphysics. It is equivalent to using the imaginary number *i* (the square root of -1) in maths: it has no reality, but it helps to solve problems. For those of a more liberal faith, there is a whole school called 'process theology, derived from this work. For me, this God postulate is an absolute

---

[7] P&R p105, 346 & 250

requirement if we are to claim any rational understanding of the world. Anyone who agrees with this must immediately start to consider the consequences in terms of how we live and the nature of our relationship to God.

Max: And for me, God will remain just that, a mere metaphysical postulate, a destructive one at that, and one we can do without.

Freya: So what's our take home message?

Orin: With our interpretation of consciousness backed by Whitehead's metaphysics, scientists can stop talking alienating nonsense about us being machines and accept the obviousness of life's conscious self-determination. With this concept accepted, anyone presenting a popular life-science program will be able to encourage our inclinations to relate and care for the natural world.

Freya: And they will be able to describe the evolution of our bodies and its emotional and conscious states through a real struggle for survival. On a broader front, the values of science will align with the values of society. This will make it harder for those with a vested interest in supporting environmentally disastrous actions to ignore the real scientific concerns being raised. The message is that science now supports our humanity rather than degrading it.

Sources and Discussion

Online, you will find an appendix with my notes related to the underlined text and a substantial list of references. Here you will also find an open access and searchable pdf version of this book with active links to the relevant section of the appendix. You can also submit your own comments and corrections.

Please go to: - http://www.originsofself.com

The Origins of Self

Subject Index

147

About the Author

Steve Brewer likes to find out how things work and fix them. This explains his interest in science, philosophy and DIY. After studying Biochemistry at London University and being awarded a Doctorate in Microbial Physiology at Bristol University, Steve pursued a career in both the UK and USA discovering and developing medicines. He now lives in Cornwall, England pursuing his interest in science and philosophy, as well as fixing things.